SpringerBriefs in Applied Sciences and Technology

PoliMI SpringerBriefs

T0214870

For further volumes:
http://www.springer.com/series/11159
http://www.polimi.it

Seçil Uğur

Wearing Embodied Emotions

A Practice Based Design Research on Wearable Technology

POLITECNICO
DI MILANO

Seçil Uğur
Dipartimento di Design
Politecnico di Milano
Milan
Italy

ISSN 2191-530X   ISSN 2191-5318   (electronic)
ISBN 978-88-470-5246-8   ISBN 978-88-470-5247-5   (eBook)
DOI 10.1007/978-88-470-5247-5
Springer Milan Heidelberg New York Dordrecht London

Library of Congress Control Number: 2013930609

Printed on acid-free paper

Springer is part of Springer Science+Business Media (www.springer.com)

Preface

Emotional expression has many crucial roles in social interaction. People rely on the emotional codes that are embodied through facial expression, intonation, gesture, or posture. There is a strong relation between the mind and the body during the embodiment of emotion. The human body is the place, where emotions are felt and expressed. Today, people are in an era of digitally mediated Human-to-Human Interaction, which cannot provide full body presence and therefore, emotions cannot be communicated completely. The lack of whole sensorial experience of the other person's presence can cause misunderstanding and emotional distance between two people. However, with the great developments in technology, today it is possible to stimulate the sensorial organs and create a multisensorial experience by enhancing the human body's role in the interaction. The digital displays have been giving their places to tangible interfaces, where the user interacts with the digital data through a tangible object. Wearing the technology is a way of creating a tangible interaction. The intimate cover of the human body, i.e., garment is the interface, where many personal traits are embodied. With the improvements in textile and electronics industry, this embodiment can be carried on a higher level, where the garments become dynamic tangible interfaces and extensions of the human body. These garments can communicate the emotions of the wearer to another person in near distance, as well as in far distance. Communicating emotions through wearable technology can introduce new interaction avenues and enhance the human body's role in mediated communication.

This book consists of a research on skin, clothes, and technology as extensions of the human body, technology-mediated emotions, and a design practice that explores the communicative level of wearable technology through turning it into a living surface, which can convert intangible data to tangible in order to provide an emotional communication. This book aims to show how research through design approach can create tools to question the Human–Technology interaction in communication context. Through creating several prototypes, the research tries to observe the user behavior toward the embodied interaction. The observations are analyzed in order to introduce new questions that can trigger designers of new technologies to create new interaction models by not only focusing on esthetical and functional issues, but also paying attention to physiological, psychological, and social aspects of the interaction. In the design practice, the Human–Technology interaction is carried into an alternative context, where technology dissolves in use and starts serving for enhancing Human–Human interaction. This book underlines the designer's role in creating new technologies, which are not anymore visible as a physical substance but integrated and embedded into everyday life garments.

Seçil Uğur

Acknowledgements

For the contribution to Social Skin prototypes, I would like to thank I. Laura Duncker who worked on the prototypes with me, and Prof. Stephan Wensveen for his supervision during the period of the project done in TU\e Wearable Senses Lab. For the technical contribution for the Skin&Bone prototype I would like to thank Martin Ouwerkerk, Florian Neveu, and TU/e Wearable Senses Lab. For the photographs of the Social Skin I would like to thank Maria Ru (photographer), Mai Men Chim (stylist), Kathleen van Walle (hair stylist), Jorien Kemerink (female model), Joost van Duppen (male model). For mechanical and technical contribution to EMbody prototypes, I would like to thank Mario Covarrubias Rodriguez, Alessandro Mansutti and KAEmart Group, Mechanical Engineering Department, Politecnico di Milano.

For the contribution to the programming part of the prototypes and his moral support, I would like to thank Volkan Yavuz. For the photographs of the EMbody protoypes, I would like to thank the models, Neslihan Ozan, and Cigdem Coroglu.

I would like to thank my tutor Prof. Raffaella Mangiarotti and my supervisors Prof. Monica Bordegoni and Marina Carulli for their great guidance in this challenging work for three years and supporting me in every step of my Ph.D. Besides, I would like to thank my research unit, Industrial Design, Engineering and Innovation—IDEA, Politecnico di Milano, and my colleagues, for their support and contributions on my Ph.D.

Contents

Chapter 1
Introduction

Abstract This chapter focuses on the research questions and the methodology that is based on a design practice. The practice-based design research addressed in this book argues that wearable technology can provide embodied emotional communication, of which the affective bandwidth level is increased by the presence of multimodal sensorial experience of emotions. This research aims to discover new ways of sensing, communicating, and experiencing emotions through wearable technology by carrying HCI (Human-to-Computer Interaction) into an alternative context, where technology dissolves in use and the interaction begins to enhance Human-to-Human interaction.

1.1 Research Questions

Emotion gives sense to people's behaviours in order to add their relationships more understanding, trust and intimacy. It is an important fact of interpersonal relationships. Nevertheless, with new communication technologies the distinction between self and society has been evolving towards a more flexible state and shifting between interpersonal to mass, private to public. While new media and communication technologies are reshaping social dynamics and privacy issues, they are introducing new interaction avenues that change traditional communication habits. The research that is addressed in this book aims to enrich the social bonds by enabling people to communicate their inner states through a tacit interaction achieved by wearable technology. This research questions *if wearable technology can be a social medium to create new social interaction.*

Over the last decades, people have become more and more aware of their wellbeing and the state of their bodies. Body sensor networks are embedded into garments and track bodily data in order to sustain healthier lives. In this new era, psychological wellbeing, such as stress control and emotion regulation have become important issues. By stimulating different sensorial channels, emotions can be induced or regulated in the body. This research aims to find new ways of evoking emotions through wearable technology in order to heighten the awareness of wearer's psychological state. This research questions *how wearable technology can awaken sensorial perception and induce emotions.*

The boundaries of art, design, fashion and technology are dissolving and creating a joint experimental area of research, which leads to product developments for the new needs of rapidly changing lifestyles. Wearable technology is one of the results of these overlapping disciplines. This research aims to explore the implementation of wearable technology in a social interaction context by applying the inter-disciplinary knowledge in a design practice. While underlining the designer's role in developing new scenarios for technologies, this research questions *how a multi-disciplinary approach can open new ways to design wearable technology.*

While textiles are soft and intimate, technology is hard and cold. This research aims to explore various ways to integrate garments and technology in order to find new interaction avenues that are more human friendly, soft and homogenous. Moreover, wearable technology is still a new phenomenon for people to accept in their daily lives, although there is a very fast adaptation to new technologies. Due to the inventions of smart textiles and soft electronics, wearable technology might move to a more aesthetic level through leaving its cyborg look behind (Seymour 2008). This research aims to be an example for designers to create socially acceptable and aesthetic forms of wearable technology. This research questions *how a designer can integrate technology and garments by fulfilling the user needs.*

Emotion has been an issue that is sometimes hidden from public view and is strongly related to privacy and intimacy. According to Klosek (2000), personal privacy has become an important issue because of the wide availability and use of electronic networks that create the ease of personal data circulation. This research tries to explore *how users react to privacy issues relating to wearable technology that expresses emotions.*

1.2 Methodology

Research through design that comprises of methods and processes from design practice has been used as a common methodology in HCI studies (Zimmerman et al. 2007). Research through design is similar to action research. Action research starts with an ambiguity or a conflicting situation that is solved through a spiral of cycles: planning, acting, observing, and reflecting in a participatory activity (Swann 2002). Flayling (1993) argues that action step continuously follows reflection step. The research addressed in this book is based on the methodology of research through design, where the design practice is the medium for answering the research questions and verifying the hypothesis. This methodology is applied through prototyping and testing wearable artefacts that are designed for communicating emotions. Dunne and Raby (2001) say that critical design makes people think. Therefore, rather than reaching to specific results, this research aims to open up new questions that can be inspiration for design of new technological artefacts

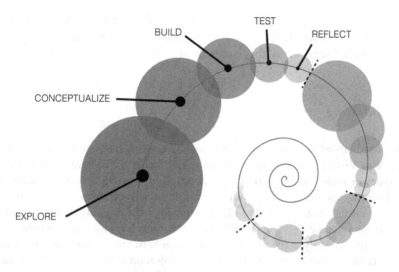

Fig. 1.1 Methodology loop that consists of five phases

and interactions. Through observing the interaction between the user and prototypes, the research overreaches the quantitative data and brings out qualitative data that can be used in order to give rise to new design approaches and practices.

The methodology is applied as a loop, of which each cycle has a sequence of steps: *exploring, conceptualizing, building, testing, observing* and *reflecting* (Fig. 1.1). The reflections on the results of each cycle give rise to new research questions that need to be explored based on a multidisciplinary literature review.

1.2.1 Exploring

Concepts of other disciplines are often used by design researchers and are discussed how they can be applied in design practice (Graver 2012). The *Exploring* step embraces a literature review of cognitive and social sciences. This literature review is addressed in Chaps. 2 and 3. Although literature research can provide valuable data about emotions and wearable technology, user surveys and questionnaires are carried out in order to acquire more data that serves for the following steps.

1.2.2 Conceptualizing

In this step a concept generation is done in order to base the practice on a context and describe whom this concept is directed at. Each concept consists of user profiles that have different motivations and needs. Therefore, in order to go deeper into each concept, different product ideas are generated. These ideas are mapped and analysed according to their feasibility and efficiency. The selected ideas can pass to a further step, where they are built as prototypes and tested by users.

1.2.3 Building

In this step, virtual and physical prototypes are constructed based on the concept generation. Virtual prototyping that is an initial model for designing the physical prototypes provides an early study to understand user perception. The findings of virtual prototypes can give directions to the physical prototypes and their developments. Virtual prototyping provides advantages on doing further modifications on the design according to obtained reflections. Besides, physical prototypes are built in order to analyse the complex interaction between the user and the prototype. This step embraces operations such as, integration of textiles and electronics, aesthetic studies, ergonomics and interaction design processes.

1.2.4 Testing and Observing

The prototypes are tested in order to explore different interaction issues in real-time context and controlled settings. Observations of user behaviours are used to guide the research and help to enhance the following cycles. This step obtains both quantitative and qualitative data that are collected and analysed in various ways. Qualitative data is gathered by observations and reports of user experiences. The observation is made by continual photographing and video recording during the test. The video recording can be useful for observing the participant in detail, but it may also infer the participant's behaviour. Thus, a verbal and written form of reporting after the test can give more privacy to the participant while experiencing the prototypes. The verbal and behavioural data of the participants are analysed for classification and interpretation. On the other hand, quantitative data is gathered by scale base questionnaires done after the user test in order to provide comparison among the prototypes and measure the various aspects of the user experience.

1.2.5 Reflecting

In the end of the cycle, reflections are done in order to trigger a new question for a new literature exploration and, therefore a new cycle. This is the end of a cycle, and at the same time the beginning of a new one. Hence, this research aims to not only find answers but also create questions that can inspire designers of new technologies.

References

A. Dunne, F. Raby, *Design noir: the secret life of electronic objects* (Birkhauser, London, 2001)

C. Frayling, Research in art and design. Royal College of Art Research Papers **1**(1), 1–5 (1993)

W. Graver, What should we expect from research through design? In: CHI 2012, ACM, Austin TX, 5–10 May 2012

J. Klosek, *Data privacy in the information age* (Quorum Books, Westport, 2000)

S. Seymour, *Fashionable Technology: The Intersection of Design, Fashion, Science, and Technology* (Springer, New York, 2008)

C. Swann, Action research and the practice of design. Design Issues **18**(1), 49–61 (2002)

J. Zimmerman, J. Forlizzi, S. Evenson, Research through design as a method for interaction design research in HCI, in *Proceedings of the Conference on Human Factors in Computing Systems* (ACM, New York, 2007), pp. 493–502

Chapter 2
Human Body's Extensions

Abstract The human body is more than a physical organism; it is a bilateral gate, where both perception and communication occurs. Throughout history, the human body has been painted, deformed and covered in order to express ideas, beliefs and ideologies. With great innovations in science and technology, the human body has been altered, transformed and extended through artefacts both perceptually and cognitively. There is a shift in technologies as desktop devices to technologies that intensely penetrate our lives and bodies. While this chapter focuses on three ways of extending the human body—skin, clothes and technology -, it argues the fact that with today's technologies these three mediums can be synthesized into one that has vanished from consciousness and serve for an embodied interaction.

While the mind and the body were always seen as separate entities in the Cartesian belief, with the discourses about phenomenology they have been considered as ambiguously connected (Merleau-Ponty 2005). While the body itself is an extension of consciousness, it can also be extended through other artefacts. The human mind and, therefore the human body are always expanded through the use of tools. Polanyi (1952) mentions this phenomenon by arguing that the tools are assimilated as parts of the human body when they are in use. Merleau-Ponty (2005) also mentions the same remark via the example of a hat, a car, or a stick. While the body is engaged with an artefact, there is no certain distinction about where the body starts and ends. The artefact can be considered as a part of the body. Heidegger explains this phenomenon with his theory of "ready to hand" (Laverty 2003). He gave an example of a hammer that disappears from consciousness when it is in use. Only when it breaks or something goes wrong, the user might realize the physical existence of the hammer, as Heidegger called "present-at-hand" (Laverty 2003). Ihde (2010), who is known for his theory of human-technology relations worked on the interaction between the human body and artefacts from an "embodiment" perspective. He explains this interaction with an example of chalk: when there is more transparency, the artefact functions in a better way (Ihde 1977). On the other hand, starting from Husserl and Heidegger, Dourish (2001) generated the idea of embodiment in technology, through integrating tangible interfaces with social computing.

S. Uğur, *Wearing Embodied Emotions*, PoliMI SpringerBriefs,
DOI: 10.1007/978-88-470-5247-5_2, © The Author(s) 2013

Skin is the first and the most intimate extension of the human body. It is a gate that continuously transmits information between inside and outside. The skin is covered by a second extension, clothes—the "second skin" of the human body. Due to the inventions of new types of textiles, such as conductive fabrics, shape-memory textiles or textile sensors, the clothes have become intelligent surfaces that can be considered as living tissues with their dynamic and smart attitudes. Besides, with emergent interface modalities, such as pervasive and ubiquitous interfaces, technology has become more and more subtle in interaction, where the human body has come into prominence. Technology has started to expand the body's sensorial capacities and become new invisible prosthesis of human body that connect it with the entire world.

2.1 Skin as an Extension

Although skin seems merely a surface that wraps the human body in order to protect the sensitive mechanism, it is the largest organ, which has many functions, such as tactile sensation, respiration and communication. This multi-tasking sur-face is a physical barrier that exists between the inner and the outer world. It is the first layer of extension, where the self starts interacting with the environment and others. Goffman (1959) defines two kinds of bodily expression: the expressions that are "given off" without control and the expressions that the body "gives" by one's own will. Hence, the skin also functions in the same manner: it is a natural display that gives the self away through changing its state according to the emotions and it is a canvas that is painted, decorated and modified in order to express personal identity. Throughout history, this important frontier of the human body has been a topic of many discourses, from art to philosophy with its para-doxical meaning that shuttles between concealing to revealing. Due its multi-tasking properties and profound meaning, it has become an inspiration for design practices in order to create new product concepts and materials.

2.1.1 Paradoxical Skin

Skin is a surface that is full of paradoxes. Taylor (1997) compares the skin's different layers by explaining them with a "hide and seek" process and argues that the skin is not just a container or something simple that covers the body; it is a paradoxical phenomenon. It is alive and at the same time dead. This contradiction comes from its Latin roots: *cutis* (living skin) and *pellis* (dead skin) (Connor 2003). Alive skin is dynamic, expressive and revealing; while the dead skin is just wrapping, covering and concealing the organs. These two opposite meanings of skin—concealing and revealing- have been paid different attentions to in different periods of history. Until the Renaissance, the human body was a closed object and the inside was a secret mechanism. The famous anatomists of the sixteenth century, such as Andrea Vesalio, Juan Valverde de Amusco, etc. pictured the mysterious inner system of the human body functioning and alive, while its skin was flatted and removed away as a cover (Bohde 2003). For instance, Michelangelo in his famous painting the *Last Judgment*, painted *St Bartholomew* as a self-portrait, whose skin was flayed and separated from its soul. This representation of skin was referring to its concealing function. These examples from sixteenth century show similar understandings of skin: without the inner system the skin was nothing, nothing but a cover. The most important function of the skin was considered as hiding the human organs. From the eighteenth to nineteenth century, the skin was given an important place within physiological and anatomical studies. The skin has become the place, where body and mind meet. From the representation of the body without skin in the early sixteenth century, to nineteenth century, the skin has started to attract attention of many discourses. In late eighteenth and early nineteenth century human skin was referred to as an interface, which functions between in and out, a transmission surface (Fend 2005). Since the eighteenth century physicians and anatomists have studied skin in a more detailed way through their microscopes and have discovered its channels that link the inner system with the outside world (Mieneke 2009). Hence, the meaning of skin has shifted from a mere cover to a functional, living organ.

Emotions and the skin are strongly connected. It is the place, where human beings experience pleasure and the pain. By the end of nineteenth century, in psychology, the human skin has been studied as a surface for emotional analysis. Skin has electrical resistance changing during emotional arousal. Duchenne (1990) induced facial expressions by using electro-shocks and recorded them to analyse the nature of human expressions. This study was taken deeper by Darwin (1872) on human and animal emotions. Moreover, Freud (1927) mentioned the importance of skin in psychology and argued that body ego rises from the experience of the skin and distinguishes the self and the other. In the same timespan, not only in psychology, but also in contemporary art the skin was manipulated as a surface, where expression happens, while until the Renaissance the skin was pictured as perfect as porcelain (Fend 2005). For example, in the painting of Edvard Munch—"The Scream"—the person's face is over deformed by the expression of fear. Francis Bacon also pictured the human body with a skin, which was blurred and transformed with the expression of fear in his painting, "Pope's screaming". These similar works of art show that the skin as a natural expression surface can go further from its limits and be a silent language of the human body. The skin transforms into new forms in order to reveal the inner state of the person.

2.1.2 Skin Adornment

While the skin exposes the inner state without self-control, it also expresses personal messages through adornment. Throughout history, skin has been continuously decorated, transformed and extended into new forms. According to Polhemus and Proctor (1978) the human body can gain a symbolic meaning by being painted, tattooed, adorned and modified. Skin adornment has a long history from tattoos, piercings to scarification. Every culture has various reasons to inscribe significant meaning on their skins (Mascia-Lees and Sharpe 1992). Human beings have adorned their flat skin with symbols to communicate status, attract attention of the opposite sex or it functions as a spiritual protection (DeMello 2007). Not only in the primitive cultures, but also in modern societies, with the emergence of tribal culture, tattoos are applied on the human skin in order to give personal messages. Body adornment has also other functions besides being a symbolic surface. For instance, the difficulties in seeing one's own body led people to paint other's bodies; and this reinforced the social bonds between the people in rural settings (Polhemus 2004).

Scarification is another type of skin adornment. This type of skin adornment is done by cuts on the skin to create permanent scars as tactile and visual patterns. Besides being a sign of status, identity and beauty, the body is scarred as a kind of protection against magic and diseases by injecting a small amount of poison (DeMello 2007). This shows that the skin is considered as a surface, which is not only between the self and the society; it is also a spiritual gate that bridges the inner and the outer world. Later, with the improvements in biomedical technologies, this type of adornment has been used by cyberpunks that have created new body styles through sub-dermal implants of laser created brands (Pitts 2003).

With the improvements in science and technology, today the permanent feature of the body adornment has turned into dynamic state and gained new dimensions. For instance, *SKIN: Tattoo* -a dynamic tattoo concept that grows with touch- shows how skin adornment can go beyond with new technologies and materials (Philips Design 2007). This project expresses the sensitivity and expressiveness of human skin in a future concept, where technology is not any more visible but functions as an invisible factor that can be placed beneath the skin. *Skinthetic* is another project, a series of design proposals suggesting how in the near future consumer branding will extend to the human body with implant technology (Allen et al. 2012). This concept demonstrates the phenomenon of skin as being an expressive surface and how it can be ironically marked through the use of technology. The skin has been adorned and marked for many reasons, and it will always be the surface, where the human body extends its personality into society. Developments in science and technology can alter skin adornment into a flexible and dynamic level, where it adds new avenues for expressing and communicating the self-identity.

2.1.3 Postmodern Skins

Besides its concealing and revealing functions, today with the influence of new technologies, skin has gained different dimensions. With postmodernism, the reality has become fluid, flexible and non-linear. In this realm, the human body is no longer seen as a solid thing, but a flexible matter, which could transform and reshape itself according to the context. The skin, which was first altered in order to be aesthetically attractive by adornment, now can be transformed by plastic surgeries. For instance, Orlan used her body as an experimental surface to give it new shapes by having several plastic surgeries through defining her body as her software (Adams and Onfray 2002). Orlan was questioning aesthetic values and tried to break them by her provocative art performances.

Due to the influence of digital technologies, the skin is not any more a boarder but a connection between the self and the outer world (Wegenstein 2006). Stelarc (1995) argues that the skin as a barrier fades away and the self can cognitively interact with the digital data. The idea of McLuhan (1964) that the media is the extension of the nervous system has become real with the emergent communication technologies, which are blurring the distinction between real and virtual. When the distinction disappears, there is no anymore a separation and difference between two entities. They become one thing. Hence, while technology is blurring the barriers, skin being considered as the most intimate boarder of the self can vanish from consciousness through enabling the human body to become more flexible, versatile, dispersed and connected.

2.1.4 Skin Metaphor in Design Practice

An object fundamentally consists of a structure and a skin that covers this structure. Skin is the surface that gives a complete and unique form to the structure and at the same time it is the place, where the user gets to interact with the object through touching. In design, there is a strong analogy between the skin of an object and the human skin due to its meaning, form and function.

In product design, the skin metaphor can be used to soften the rigid lines into organic shapes that are more familiar to the human body in order to create comfortable and pleasurable products. In 1969, Italian designer Gaetano Pesce designed *La Mamma* that inflates and transforms into a woman shape sofa that is connected to a small baby ottoman. This is an intelligent use of a new material that shows how skin is represented in two states, death without air, and alive like a human body, full of air. In the 1970s, new materials such as latex, vinyl rubber, and resin, led designers to invent new sensual forms (Lupton 2002). Mario Bellini's typewriter for Olivetti had a skin colour rubber surface and was designed in order to create a skin like soft sensation as a different user experience. GINA, a concept car of BMW has an exterior surface, made entirely out of textile fabric that covers the metal frame of the car (BMW 2008). When the frame moves by its electro-hydraulic system, the fabric skin stretches and transforms. Lupton (2002) introduced a very profound work that presents examples from product, furniture, fashion design, architecture, and media that are created with special skin concepts and addresses the issue of skin as metaphor for design practices.

While designers were using skin's elasticity and tactility as an inspiration in their works, the innovations in material science have carried these attempts one-step further, to an intelligent state. As mentioned before, skin has two layers: one is death, and the other one is alive. While the dead skin protects and covers the organs, living skin has the sensors and the nerve endings that detect tactile stimuli, changes colour, self-heals and deforms. Today materials can act like a living tissue with their intelligent features. The materials that are reactive to outside stimuli are used in various product solutions. For instance, Nokia (2008) created *Morph Design Concept* that is a mobile device, which changes shape, creates self-cleaning surfaces and measures the environmental parameters in order to analyze the quality of the surroundings through nano-technology. Nendo (2006) designed *Hanabi Lamp* that is made of shape memory alloy and changes shape when the temperature of the lamp increases. The intelligent cover of the light can open like a flower and gives more light over time. On the other hand, the *History TableCloth* is a flexible screen- printed cloth with electroluminescent material that glows slowly, when an object is put on it and disappears when it is removed (Gaver et al., 2006).

Both skin and buildings protect and shelter the human body. Buildings are the "collective epidermis" that covers more than one person (Krueger 1994). Due to the improvements in material science, the skin metaphor has become widespread among architecture. Buildings with static covers have shifted to have flexible, versatile and dynamic membranes (Wigginton and Harris 2002). Textiles gained extreme features and became stronger, lighter and smarter (McQuaid 2005). These new textiles can be used in architecture in order to provide thermal insulation, light transmission and noise reduction.

Skin- the first frontier of the human body- has been an important analogy for designers and led them to shift their attention from the objects with mere surfaces, to the objects with intelligent skins. Hence, designing an object is not any more just a problem of aesthetics or function, but also involves a wider approach that focuses on user experience. The designer should be the creator of new experiences and interaction scenarios, in which the user is engaged with the artifact.

2.2 Clothes as Extension

2.2.1 Philosophy of Clothes

After skin, clothes are the most intimate surfaces between our surroundings and our bodies. One of the primary roles of clothing is to cover and to keep the wearer comfortable and safe. Although the first human beings covered themselves to protect their bodies, by time covering has gained other functions such as trans-mitting messages about the identity, social statue, ethnic class, religion, gender, and mood. Human beings are the only living creatures who are dressed. Clothes are social communication mediums that are visual signs and representations of social relationships (Carlyle 1831). People dress their body to be socially acceptable in the context, where they live. There are different dress codes for different contexts, cultures and circumstances. In order to avoid social critiques and to be accepted, people try to dress themselves suitable (Bell 1976). People may put clothes on in order to make divisions in society and communicate these differences. For instance, uniforms were created to distinct the affiliations of people at working environments. Eco (1986) suggests that dresses have a language, through which people speak.

Another function of clothing is to cover the vulnerability of the body. Clothes are used for concealing the sexual organs and the intimate parts of the human body. When the body is revealed more than the norms of the society, this can create shame. However, clothes can simultaneously both hide and draw attention to the body. Bell (1976) describes that a wrapped object can be perceived more beautiful and attractive than an uncovered object. A cover can create curiosity that increases the imagination of what is covered.

Douglas (2002) defines two kinds of body: social and physical. While the social body represents the reflection of society, the physical body stands for the experience how the person physically feels in his/her own body. Entwistle (2000) also uses a similar taxonomy and mentions two kinds of function of clothes: "intimate experience" and "public representation". The intimate experience of clothes embraces physiological factors, such as pressure, dryness, softness, roughness, smoothness due to the close interaction with human skin. Zimniewska and Krucinska (2010) found out that textiles with different fibres could be perceived as pleasurable or disturbing. Not only tactile sensation, but also auditory and visual factors are important for the intimate experience of clothes. For instance, Cho et al. (2005) evaluated the sound of polyester wrapped knitted fabrics by using a psycho-physiological technique and found that the rustling sound of the fabric could affect psychological and physiological responses through evoking pleasant and unpleasant sensations. Auditory experience can affect the physiology of the wearer, and therefore it is psychology, too. Besides, colour associations can also cause different negotiations, and different perceptions of one's own body; therefore it affects the wearer's psychology. Movement is also another important factor of the intimate experience of clothes. The dress can change the way, in which the body moves. The dress can restrict the body or create ease, while the body is performing. Therefore, the elasticity and how the cuts of the dress are placed on the body are crucial for creating a comfortable garment. Many sports-wears are designed according to placement of muscles in the body in order to enhance the wearer's movement. The body has areas that have different thermoregulation, sensitivity or movement capacities. While designing a dress those properties of the body should be taken into consideration to obtain maximum comfort and ease to move.

Kansei classification that is used for analyzing user's experience in terms of emotional level (Nagamachi 1995, 2002) can be applied also in research on clothing. For instance, Shaari et al. (2003) defined four female Kansei classifications for traditional clothes: somatic that creates tactile sensation on the skin, cognitive that causes emotional reactions through personal experience, pleasing that is related to aesthetics of the material and shape and social that becomes a social symbol communicating social affiliation, identity or status.

According to Carlyle (1831) besides protection and decency the first purpose of clothes is ornament. People do not just cover their bodies; they dress their bodies to be aesthetically attractive and beautiful. The fashion phenomenon started with the difference between covering and dressing the body. Flügel (1930) used Freud's model to explain how people are born with a narcissistic characteristic that admires its own body and shows it to others in order to share this admiration. People dress their bodies to attract others and, therefore this can create a satisfaction of being liked. Not only the shape, tactility and colour of the dress please the wearer, but also how the wearer is perceived as a body image is important. For instance, a dress may feel uncomfortable, but the wearer puts it on to be perceived beautiful by the audiences. *Body Cathexis* is the satisfaction with the appearance of the body (LaBat and DeLong 1990). The body is caged beneath the clothes in order to exhibit a desirable shape for the self and others. A dress sometimes can become a camouflage of the undesired parts or a prosthesis that gives a new form to the human body. The garments can reshape the body according to the will of the wearer. For instance, the corset or bras are the most important women's underwear that moulds the body in order to give an aesthetic appeal to the wearer. This idea of altering the body like prosthesis has become a way to create new human figures in fashion. In the last decades, some fashion designers have worked on the issue of garments as prosthesis that are attached to the human body. For instance, Lucy and Bart (2008) created futuristic human shapes, by using a prosthetic approach for human enhancement. Besides, Burke (2012) designed sculptural forms around the shape of the human body inspired from prosthetics and medical tools. Those examples show how clothes can become prosthesis of the human body, which alters its limits by creating new body figures.

2.2.2 Clothes as Second Skin

According to Fiorani (2010), wearing clothes is like wearing another body on top of the natural body. Clothes gain meaning and life, while they are worn on the body, and they become empty shells without life, when they are separated from it. Clothes can move, curl and change its shape according to body movements. For instance, drapery in Roman clothing emphasizes the movement and reveals the body as attractive and full of life (Hollander 1993). The folds, one after another give an infinity perception (Delueze 1992). The fold as activator of movement can give life to the garment. In many spiritual dance rituals, the dance costume expresses spirituality with its movements. The body becomes fluid with the dress and represents the soul, which can be considered as the living part of the body. For instance, in dervish dance or tanoura dance, dancers wear special skirts that move up when the body starts whirling. Many dance costumes are designed as extensions of the dancer's movement. In the beginning of the twentieth century, Loie Fuller developed her own costume design by combining her choreography with silk costumes illuminated by multi-coloured lighting and her dance creates drapery that becomes continuation of her body, while she is swirling (Doy 2002).

Many artists and fashion designers have used the "second skin" metaphor to express the intimate interaction of clothes and the body. Turner (2008) defines the human body as "dressed body". The human body gives life and fullness to the dress (Entwistle 2001). The dress becomes alive with the body's breath. Bell (1976) argues that the clothes are indifferent from the natural body, as being extensions of the soul.

Elsa Schiaparelli designed The *Skeleton Dress* that had embossed lines like the chest bones as if the dress represented the human skin wrapping the inner structure. Nanni Strada designed *la Pelle* (the Skin) that was a tubular knitted dress covering the human body as a second skin. *Skin-maternity clothing* by Marisol Rodriguez also functions as a second skin that deforms and enlarges itself during the pregnancy (Rodriguez 2009). These examples show that the paradoxical substance of skin, which shifts continuously from concealing to revealing, is often used as a metaphor in the fashion field. On the other hand, skin can become a biomimetic metaphor for fashion. There are growing attempts to merge biology with design due to the improvements in tissue engineering. In her Bio Couture project, Suzanne Lee uses yeast and bacteria to grow fabric: cultivating the material in a sweet tea solution (Verhoog 2007). The *(In)visible Membrane* project by Sanja Baeumel consists of self-growing wearables that react to body temperature (Seymour 2010). The Tissue Culture & Art's *Victimless Leather* -semi-living garment- is cultivated via growing living tissue and transforms into a leather like material (TCA 2004).

2.2.3 Smart Clothes

The way people dress changes according to cultural, economic, social and technological evolutions. Garments are the mirrors of society; they reflect not only personal identity, but also a general picture of the society that is highly influenced by technology. In the 1920s, Giacomo Balla suggested a manifestation on "il vestito antineutrale" (anti-neutral clothing) that described future clothes as dynamic surfaces, which could transform into new shapes, colours and styles according to the needs of the wearer (Woodham 1997). His manifestation was a forecast, which after a century became real. While technology has entered into almost every object, it has started to become embedded also into garments. Technology, in an invisible way can turn the garments into smart skins that allow the wearer to be more connected, protected, sensitive and expressive. Today, clothing is passing into a new dimension, where it is not any more just a passive surface, but behaving as a living, sensing and transforming interface. Due to the integration of electronics and textiles, smart clothing has emerged in order to track body signals, stimulate sensory experience and establish connection between bodies. With smart clothing, information is continuously transmitted through the human body, from inside to outside and vice versa. Seymour (2010) mentions smart textiles and garments as a second epidermis of the human body that can stimulate senses and create dynamic attitudes as "animated canvas".

Technology and garments have always been intimately related. Jacquard's loom that is used for weaving textile is a primary version of digital devices (Ryan 2008). The way, in which textiles are built, is very similar to how computers work. Therefore, embedding technology into fabric is a kind of a natural evolution. Microelectronics have become small and inexpensive enough to be embedded in the daily clothes. Moreover, with wireless technologies, intelligent garments can be connected and share data. Although the first application of smart clothing was done in the military field in order to increase security, today there are many application areas, such as health, wellbeing, sports, leisure, and communication. Ariyatum et al. (2005) mentions three phases of the design evolution of smart clothing: smart clothing in 1980–1997 that is driven by technology, smart clothing in 1998–2000 that is influenced by fashion and the textile industry, and smart clothing in 2001–2004 that is oriented to the market. In the first period of smart clothing, the innovations came mostly from computer science as a top-down approach, which aimed to put computers into clothing (Lee 2005). This period is known with the image of Steve Mann's wearable computer projects in MIT (Mann 1996). Later, the concept of wearing rigid technology on the body shifted into a more flexible form thanks to the smart textiles and soft electronics. One of the earliest electronic garments, in 1996 the *Smart Shirt* was created by using optical and electrically conductive fibres that monitor the wearer's vital signs, such as heart rate, EKG, respiration and temperature (Bowie 2002). In 1999, MIT Media Lab designed the *Smart Vest* that was soft and comfortable with the purpose of integrating technology and the human body without interfering the image of daily clothing (Schwartz and Pentland 1997).

In the second period of the smart clothing evolution, fashion and textile industries got involved in producing smart clothes through a bottom-up approach that aims not to put the computers into clothing, instead to make clothes that have special functions (Lee 2005). The projects were done with collaborations of different fields: fashion, textile, and engineering. With the invention of new textiles, fashion designers became more innovative and created new looks and functions under the name of smart clothing. In 1997, a fashion show dedicated to smart clothes was done by Creapôle Ecole de Création and Alex Pentland (Ryan 2008). In the same period, Meggie Orth at the MIT Media Lab developed new methods for embedding electronic circuits into fabric and used conductive textiles in order to create soft electronics (Parsons 1999). But still these garments were not ready to be in the market.

From 2001 to today, smart clothes have found their places in the market. The Levis *ICD+ jacket* was created in collaboration with Levis and the Philips Research Laboratories, considered as a commercial wearable technology product (Quinn 2010). Brands such as Nike and Adidas have started to introduce smart products in their fashion lines. *Adidas 1* is one of the earliest examples of commercial sport products that merge technology with shoes (Fiorani 2010). Besides commercial market, DIY (Do It Yourself) smart clothing has dramatically increased in recent years. With the rise of open source knowledge and technologies, people have started to build their own intelligent clothes. New tools and methods were invented in order to create DIY kits. For instance, the *LilyPad-Arduino Kit* allows users to build soft wearables by sewing microcontrollers, sensors and actuator modules together with conductive threads onto garments (Buechley et al. 2008). Perner-Wilson and Satomi (2009) introduced various techniques on how to create DIY textile sensors, such as pressure sensors, potentiometers, tilt and stroke sensors that are handmade with low-cost and low-tech techniques.

Nowadays, the term, smart is used in many discourses. Smart means being able to sense the stimuli and react or adapt accordingly (Baurley 2003). A particular textile can make the garment smart by itself, or normal cotton can be turned into a reactive surface by embedding microelectronics or activators. Smart textiles can be classified in three categories: *passive smart textiles*, that can sense the environmental conditions or stimulus, *active smart textiles* that can sense and react to the stimuli and *very smart textiles* that can sense, react and adopt themselves to environmental conditions or stimuli (Tao 2001). Smart textiles can include input readers of the smart system, such as switches or sensors, output displays that react to the input in various sensory modalities, power supply, communication elements and data processing.

2.2.3.1 Input Readers

A smart garment can be activated by various inputs: such as the wearer's command, environmental stimuli or bodily signals. Textile buttons and flexible switches are one of the actuators that can operate and command the smart system. There are different ways to create textile buttons for smart clothing. One of the fabric switches is made through sewing conductive and nonconductive fabric together insulating and protecting themselves from each other with a soft, thick and holey fabric that allows them to contact when pressed (Post and Orth 1997). These switches can be easily soldered on a textile circuitry that connects to a microprocessor. On the other hand, there is also another type of textile keyboard that is made by embroidered or silkscreened electrodes, in which the conductance is increased when pressed (Post and Orth 1997). Textile buttons on commercial products are mostly used for creating ease to control mobile devices carried on the body. Soft keyboards and switches produced by *Fibretronic* are made of textile and polymer based buttons that can control wearable accessories (Fibretronic 2013). By stitching and bonding techniques these textile buttons can be attached on normal clothing in order to control the smart systems. On the other hand, QTC^{TM} *Material* that is a composite material made from conductive filler particles combined with an elastomeric binder can shift into a conductor when pressed (Peratech 2013).

Besides textile buttons, various types of fibres, such as fibre optics or conductive yarns can be woven and used as textile sensors to measure the body's physiological data. Many investigations were done to create textile sensors in order to provide ease in body tracking mostly for medical and sport applications. *Textrodes* were invented through knitting stainless steel fibres in order to measure ECG (Catrysse et al. 2004). These sensors can be used to monitor the performance of the wearer, in sport activities. Today in the market, various brands sell their wearable sensors that can be used by non-professional users for daily sport activities. Heart rate sensors are generally attached to a flexible belt that is worn on the chest area. This type of textile sensors can be embedded directly on the garment. For instance, *Numetrex* is seamless sportive apparel that is knitted with heart rate sensors (Textronics 2013). The textile sensors maintain contact with the body, sensing the wearer's heart rate and relaying it to a transmitter that is placed in a pocket in the front of the garment. The heart rate is read through a watch that displays the wearer's biological data. Adidas also has a similar application that is called *miCoach* that instantly transmits the heart rate data to a smart phone and wirelessly uploads it to an online platform in order to track the overall performance (Adidas 2012).

Body temperature can also be measured by a knitted temperature sensor (KTS) that is constructed by merging fine metal wire with knitted structure (Husain and Dias 2009). On the other hand, piezo-resistive sensors can be used in order to monitor bodily data, such as breathing or motion of the joints. Breathing rates can be measured by detecting the expansion and contraction of the lungs. Fabric sensors that are made of polypyrrole (PPy) correspond stretching by increased conductivity and can measure breathing rates (Brady et al. 2007).

2.2.3.2 Output Displays

Clothes with their intelligent features can become dynamic displays that communicate with the wearer in various sensorial modalities with various outputs: visual, auditory and tactile outputs.

Visual Outputs

Clothes are visual tools for self-expression and this function can be carried one level further with smart clothing. Smart clothes can change form, colour or emit light in order to create visual outputs on the wearer. Visually dynamic garments cannot only be aesthetically advanced interfaces, but also are functional to communicate crucial information to its wearer or to others.

Moving garments can be activated by shape-memory metals (Nitinol) or mechanical actuators. The construction of the garment is important to create movement and shape change. A flat textile can have a dynamic attitude with a simple fold, support or a flexible structure. Shape Memory Alloys (SMAs), mostly made by Nickel-titanium and copper, can be plastically deformed and return to its original shape with temperature change (Lane and Craig 2003). These alloys can be embedded into textiles or woven as fibres in order to make the garment move under set conditions. Due to the fact that some types of alloys are activated under high electrical currents and temperatures, isolation of the material is crucial to prevent body contact. Shape memory textiles can be controlled by microprocessors and programmed to transform under required circumstances. Corpo Nove used thermal shape memory metals that contain 45 % Titanium for creating a shirt, called *Oricalco* that is programmed to shorten when the room temperature increases (Kaur and Gale 2004). Leenders's (2013) *Moving Textiles* and XSlabs's *Kukkia Vilkas* and *Scorpions* (Berzowska 2010) projects were made with this innovative material in order to create moving garments. There are different techniques to integrate shape memory materials into garments: weaving\knitting, stitching and felting. SMAs are not soft enough to be knitted in the machinery; hence hand knitting or special knitting process should be used in order to create textile structures. Boussu et al. (2002) made some experiments based on weaving SMA wires of Nitinol mesh. Shape Memory Polymer (SMP) has the same effect as the Ni–Ti alloys, but due to fact that it is made of polymers, it is more adaptable to be woven as textiles. Due to the fact that SMAs need high temperatures to be set, it is challenging to find accompanying materials in order to weave as a hybrid textile. On the other hand, stitching is another method for embedding SMA wires into textiles. The fabric should have high flame resistance or the wires should be isolated in order to prevent burn out. Natural wool is highly flame resistant; therefore it can be used with SMAs that reach high temperatures when activated. Besides, knitting\weaving and stitching, traditional felting technique can be used in order to embed the SMAs into wool. The wires can be covered with the wool and hidden inside the felt structure. Wearable technology designers commonly use this technique in order to create moving garments. *Sprout I/O* is a haptic interface that was built from a SMA and felt composite by MIT Media Lab. (Coelho and Maes 2008). Berzowska and Coelho (2005) experimented with heavy hand-made felt by hand stitching and felting Nitinol wires into the protoypes in order to make them move.

On the other hand, traditional textile folding techniques, such as pleats can be used in order to create moving textile structures. Based on origami concepts, Miyake (2012) created a clothing line, *132 5. ISSEY MIYAKE*, which unfolds into shirts, skirts, pants and dresses. Although it does not contain any mechanism or electronics, it shows how textile can extend into a three dimensional space. Fashion designer, Hussein Chalayan is one of the pioneers of moving garments that are activated by various mechanisms hidden under clothes (Evans et al. 2005). He animated his collections with an invisible power, technology. By adding a mechanical system, designer Kim (2012) created a *Kinetic Mechanical Skirt* that consists of triangular grids activated by a mechanical system. *Walking City* is a garment that can react to sound or movement and move its textile folds with the help of hidden pneumatic pumps (Gao 2012). With shape changing smart garments, the human body can obtain a dynamic silhouette that changes according to stimuli, and transforms into different shapes.

Beside their shape-changing attitude, smart garments can be visual output interfaces that can change colour. There are four kinds of colour-changing systems; thermochromic, which changes colour in response to temperature, photochromatic, which changes colour in response to UV light (Lacasse and Baumann 2003), hydrochromic, which changes colour in contact with water, and piezochromic, which changes colour in response to pressure (Berzowska and Coelho 2005). Colour changing inks can be applied on the textile with offset lithography, flexography, gravure, and screen-printing or digital printing (Berzowska and Coelho 2005). *Transitional Stripes* project was made through applying liquid crystal thermo-chromic dyes on the textile (Robertson 2009). *Shimmering Flower* is another project that was constructed with conductive yarns and thermo-chromic ink together with electronic components in order to show how technology can enter into textiles in an aesthetic way and adds dynamic attitudes (Berzowska 2004). Smart garments that can change colour can be used as visual output interfaces in order to communicate bodily or environmental data in an aesthetical and subtle way.

Light emitting garments are also used in order to give visual outputs. Lighting effect in textiles can be created by using optical fibers that light up with external light, electroluminescent material that emits light with electrical current (Worbin 2006) or photo-luminescent materials that emit light in dark. Electro-luminescent wires can be embedded into garments in order to create electrically activated lighting interfaces. *Luminex* that emits light through woven fibre optics can be used as a light emitting fabric on the garments (Luminex® 2012). On the other hand, Light Emitting Diodes (LEDs) can be integrated in textile in order to create lighting textiles. *Lumalive* from Philips contain LEDs that display full colour moving images on clothing (Klink 2006). Flexible Organic Light Emitting Diodes (OLEDs) are made of an organic emissive electroluminescent layer that emits light and enables a matrix of pixels in a higher resolution (Harper et al. 2008). *iLuminate* system is a wearable, wireless lighting system that is constructed with EL (Electroluminescent) wires in order to create costumes for special performances (iLuminate™ 2012). These textile displays can turn the body into a dynamic surface, as a new way of self-expression. Personal communication can go further to a dynamic and instantly updatable level by animated visuals on the body with light emitting garments.

Tactile Outputs

Tactility is an important feature of garments due to their intimate interaction with the skin. This interaction can go beyond to a dynamic and programmable level by embedding haptic technologies into garments. Since haptic technologies have gained importance in human–computer interaction, they also entered to smart garments. Tactile smart garments can produce tactile outputs that can be felt in various forms, such as temperature, pressure or vibration. Vibration can be obtained by vibration motors that are small and strong enough to create various types of vibrations. Vibro-tactile wearables can be used in virtual reality in order to increase tactile experience (Lindeman et al. 2006) or in mobile navigation systems that direct people with tactile signals (Tan et al. 2003). On the other hand, heating coils or conductive threads can be embedded as micro heaters into the garments in order to create temperature augmentation. For instance, *WarmX* is a special underwear that consists of silver plated thread of polyamide, which is knitted into the fabric and heats various zones in the body via current flows (WarmX 2012). Texture is another type of tactile property of textiles. Therefore, texture changes can be done through activating the textile with SMAs or piezo-electronics. Electro-active polymer can also be embedded into garments in order to create tactile stimulation without additional electro-mechanical systems (Koo et al. 2008).

Auditory Outputs
Sound is another type of output that a smart garment can produce. Auditory features can be applied on garments by piezo-buzzers, micro-speakers or embroidering coils. On the other hand, sonic fabrics are available to be used as auditory output interfaces. *Sonic Fabric* is a special textile that is woven from 50 % recycled audiocassette tape and 50 % polyester and can create sound when brought in contact with a tape head (Brower et al. 2005). *Accouphène* is a jacket that is decorated with thirteen soft speakers created by embroidering decorative coils of conductive yarn (Berzowska 2010). When the stitched magnets on the jacket come into contact with the coils, they generate sounds of different notes.

2.2.3.3 Power Supply in Smart Garments

Energy is an important element that gives life to smart garments. Due to the fact that a smart garment needs to be light and comfortable, the power source should be small, portable and sustainable. Power supply elements are becoming smaller, rechargeable, flexible, water-resistant, and can be produced with lower costs (Hahn and Reichl 1999). Flexible and stretchable film batteries that are made from Li battery materials can be integrated into textiles through weaving techniques (Liu et al. 2012). On the other hand, the human body is always on the move and mechanical energy or body heat can be used in order to harvest energy for smart garments. Piezoelectric fibers can be woven with conductive fibers as electrodes in order to generate energy through mechanical strain or vibration (Vatansever et al. 2011). Solar energy is another type of energy supply for smart garments. By using photovoltaic films embedded on garments, it is possible to create energy that powers the smart system. With this new material solar panels become flexible, soft and suitable to be embedded into garments.

2.2.3.4 Communication in Smart Garments

In smart garments there are two types of communication: short range and long range (Tao 2005). Wired short-range communication is done in between components of smart garments mostly through conductive yarns. Conductive textiles can also exist as yarns, fibres or threads. There are various types and diameters of conductive thread available on the market and are used in order to create short-range communication between the components of smart garments, such as power source, microprocessor, sensors, etc. Conductive fibres can be processed on ordinary textile machinery or they can be embroidered in an aesthetical way to create circuits on the textile. On the other hand, conductive inks can also be digitally printed on the flexible surfaces in order to create textile circuitry (Meoli and May-Plumlee 2002). Wireless short-range communication is done between the smart garment and a nearby device through wireless communication technologies, such as Infrared or Bluetooth technologies. Wireless long-range communication is done in order to sustain long distance connections. By using wireless long-range devices, smart garments can receive and send messages to others through an Internet connection.

2.2.3.5 Processors in Smart Garments

Data processing can be done in or outside of the smart garment. Due to the fact that processors do not have flexible properties, they cannot fit on the garment easily. *Lily Pad* microprocessor can be stitched easily with conductive yarns on the smart garments (Buechley et al. 2008). This allows designers to create decorative textile circuitry on smart garments. Data processing can also be done through an external device, such as smart phone or computer that is wirelessly connected to the garment. The input readers send data to an external or internal processor that continuously processes this data and activates the output displays.

2.3 Technology as an Extension

While the sensory organs are extended through new media and technologies (McLuhan 1964), the limits of the human body have become ambiguous. With the influence of digital technologies, the physical and cognitive substance of human body has gained new dimensions. The interfaces, in which the user reaches to digital data, are not anymore rigid and stable, but ubiquitous and pervasive (Harper et al. 2008). Due to the miniaturization of technology and the ability of being connected, ubiquitous computing that is ambiguously interpenetrated into every-day life (Weiser 1991) has triggered new modalities of human-technology relations. Ubiquitous computing enables users to access information in a natural way in their own environment rather than forcing them to enter into a digital world (York and Pendharkar 2004). Regarding the human body's role in ubiquitous computing, the human body becomes the central element of HCI (Human–Computer Interaction) through moving away from the desktop-based interaction towards mobile and wearable applications. Rather than having static screens to interact with, ubiquitous computing enables the human body access to the digital world with a whole body engagement.

On the other hand, Tangible Interfaces that form a bridge between the abstract digital world and objects in the physical world (Ishii and Ullmer 1997) have started to change the notion of interaction with technology. With social computing that provides social connectedness through computational systems, tangible interfaces have gained a new dimension that is called "embodied interaction" (Dourish 2001).

2.3.1 Digital Extensions

Virtual-reality has risen as a way of self-extension, where the virtual and real overlaps in cyberspace (Gibson 1984). People seek to project themselves in virtual reality with their virtual personalities as they do in the real life with their bodies, clothes and objects. Sherman and Judkins (1992) describe cyberspace as a place, where the impossible things can be created with the virtual tools. It can be seen as a parallel life that everything is possible, and the ruler is the self. People can have virtual bodies with different characteristics, even with different genders (Featherstone and Burrows 1995). People see themselves extended in a virtual world, which does not exist physically. Turkle (1997) gives "Alice stepping through the looking glass" example to describe the world, which is lived on the screen. In cyberspace, people are able to have multiple bodies, create virtual communities and meet people that they would never meet physically.

The human senses play important roles in perceiving the world. However, due to the increasing usage of digital technologies and virtual-reality, people have predominantly been using their visual sense more than other sensory modalities. With the use of virtual technologies, the way in which people perceive reality -the existing matter that people can see, feel, touch and react- has changed. Is a seen object on the computer screen real? What makes it real, being touched, seen or heard? These questions that have emerged due to the increasing use of digital technologies and virtual platforms have pushed designers and scientists to create new solutions in order to bridge the physical and the digital worlds through putting emphasis on the human body and extending its sensorial capacities.

2.3.2 Physical Extensions

Besides having virtual extensions, the human body can be extended physically through technologies that can be: prosthetic, handheld, wearable or implanted (Fig. 2.1).

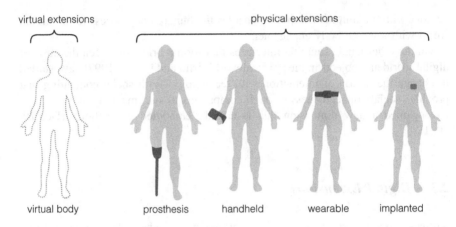

Fig. 2.1 Human body's extensions

2.3.2.1 Prosthetic

Throughout history prosthesis have been used in order to recover the missed parts of the body by adding an artificial one. In order to function in wholeness, the human body should perceive the prosthesis as a part of it. People can experience ownership toward an object extraneous to their biological bodies. To prove this, the Rubber Hand Illusion (RHI) test was done by simultaneously stroking an artificial hand that was placed in front of the participants and the real hand that was hidden behind a screen (Botvinick and Cohen 1998). The test shows that participants emotionally reacted when the rubber hand was touched and it was felt as if the rubber hand were their own hand (Botvinick and Cohen 1998). This study of the rubber hand is related to the concept of extended physiological proprioception that was pioneered by Simpson (1974). When using a tool such as a stick, perceptual experience can be transferred to the end of the tool. This shows that the body can be extended with a physical matter, which can be perceived as a part of the human body. Rebeca Horn, who is one of the pioneers of body extension art pieces, such as Feather Fingers and Finger Glove Horns, shows that the artefacts can become prosthesis of the human body, where the senses are extended to the edge (Roustayi 1989). By convincing the mind through visual and tactile signals, she addresses the issue of the linkage between body and mind and how perception can change through bodily extensions.

When technology is replaced with an organ, it becomes prostheses. Many scholars from cyborg technology use the term prostheses often as a synonym for body-machine interaction (Ott 2002). Cyborg -the symbiotic fusion of human and machine- was first coined in 1960 by Clynes and Kline (1960). Haraway (2008) claims that technologies are not mediations in-between people and the world; rather they are "organs" of the human body. Stelarc in his project *Third Hand* created an extension of him that can perform as an extra hand in the virtual space. In his projects the human body is an experimental space to understand its limitations (Stelarc 1995). While technology becomes prosthesis, it also merges with the human body by creating a symbiosis. Licklider (1960) mentioned that the symbiosis of the human brain and the computer is a kind of cognitive prosthesis. On the other hand, (Longo 2003) argues that "homo technologicus" that is a symbiosis of biological and a technological being is not a mere integration of the human body and technology, but it is rather a transformation done by technology. This symbiotic relation has been carried into a new dimension, where technology is not visible anymore as a physical prosthesis that augments the existing organs, but functions as an invisible prosthesis of the human mind through penetrating inside daily objects, clothes, surroundings and even under the skin.

2.3.2.2 Handheld

Hands are the most used parts of the body that serve for interacting with an object. Handheld devices, such as smart phones or tablet PCs enable people to carry gigabytes of data and connect with people whenever it is required. Due to the fact that hands are strongly related to tactile sensations, haptic technologies that apply forces, vibrations, and motions (Robles 2009) have been used in handheld devices. Communication technologies that so far have used visual and auditory sensory channels have entered into a new realm, where tactility has become more important than ever before. On the other hand, hands are mostly used in order to make gestures. Therefore, handheld technologies can be used in order to create gesture-based interaction. For instance, the Nintendo-Wii brings the body into interaction whilst playing the video game by a handheld console that can detect the players' gestures and motion (Schlömer et al. 2008). X-box + Kinect brought this interaction one level further and created a hand-free system that does not need a device anymore, but the human body is tracked and becomes the controller of the game (Microsoft 2013).

2.3.2.3 Wearable

Besides handheld devices, wearable technology is a new way of interaction, in which technology is worn on the body in order to provide easy interaction. Wearable technology is an attempt to free the body and allows the body to move independently whilst being connected with a technological system. In twentieth century with the improvements in computer science, people have started to place technology on the human body. In the late 1960s, Thorp (1969) designed a wearable computer to predict roulette wheels. After the 1980s, Mann (1996) opened a new era in wearable computing with his various experiments on mounting computers on the human body. Mann (1996) observed that people found the combination of human and machine bizarre. This strange image of the integration of human and machine passed to an advanced level with inventions of the electro-textiles. This invention opened a new phase in wearable technology history. It became related to fashion, rather than computer science. The fashion industry is constructed by textiles, cuts, and styles and is strongly related to aesthetics. On the other hand, technology is seen as functional tool that works with bits and digits. Integration of these two different worlds is a challenge for designers of wearable technologies. Not only aesthetical, but also the social level of integration is important. Wearable technology should be in line with social norms in order not to create complications in social interaction (Edwards 2003). Due to the fact that the integration of two different substances -technology that is artificial and the human body that is biological- is a complex process, designers should create new frameworks in order to design wearable technologies with consideration of multiple aspects (physical, psychological and social). A design-oriented approach can open up new application fields for wearable technology. For instance, first attempts have drawn a lot of attention from application areas, such as military, health and sports, where there is a need for continuous body tracking. Today, wearable technologies are used also for enhancing the emotional wellbeing that is an emerging field of research. This shift shows that the more design is involved, the more it can answer the user needs that are not only task oriented, but also related to psychological and social aspects.

2.3.2.4 Implanted

Implants refer to technologies inserted into the human body by medical surgery. For instance, RFID chips are one of the technologies that have become small enough to be implanted under the skin. Although this could alter the body, due to the fact that it requires special implementations, it is hard to use in daily life. In 1997 artist Eduardo Kac implanted a microchip in his body in order to express technology's invasive movement towards the human body (Clark 2003). In 2002, Warwick et al. (2003) inserted electrodes under the skin in order to measure neurological signals and stimulate the nerve fibres. Bilal (2013), a performance artist, implanted a small digital camera into the back of his head and streamed live records to his website.

Although implanted technologies can open many opportunities and lead to new applications, people might find it hard to accept the insertion of an artificial device under their skin. However, with the new technological improvements, the implantation can be done easily and with less risk. Regarding the acceptance of implant technologies, Sugiyama and Katz have done a survey asking the participants what function they would prefer to implant in their bodies and found that enhancing aesthetical appearance is more preferable than location monitoring, social contact and entertainment (Katz 2003). This survey shows that people can use extreme technologies to enhance their body, not only in functional manners, but also mostly for having aesthetically pleasant bodies.

References

P. Adams, M. Onfray, *Orlan: This Is My Body, This Is My Software* (Black Dog Publishing, London, 2002)

Adidas (2012), Micoach connect heart rate monitor for Iphone. http://www.adidas.com/us/product/mi-coach-heart-rate-monitor-for-ipod-iphone/SB491. Accessed 05 Jan 2013

P. Allen, C.R. Allen, Knowear (2012), http://knowear.net/mydocs/KnoWear_descriptions.pdf. Accessed 05 Jan 2013

B. Ariyatum, R. Holland, D. Harrison, T. Kazi, The future design direction of smart clothing development. J. Text. Inst. **96**(4), 199–210 (2005)

S. Baurley, Interactive and Experiential Design in Smart Textile Products and Applications. in *Proceeding of International Textile Design and Engineering Conference (INTEDEC): Fibrous Assemblies at the Design and Engineering Interface*, Edinburgh, 22–24 Sept 2003

Q. Bell, *On Human Finery* (Schocken Books, New York, 1976)

J. Berzowska, Very slowly animating textiles: shimmering flower. in *SIGGRAPH '04: ACM SIGGRAPH 2004 Sketches*, New York (2004)

J. Berzowska, M. Coelho, Kukkia and Vilkas: Kinetic Electronic Garments. in *ISWC'05 Ninth IEEE International Symposium on Wearable Computers*, Osaka, 18–21 Oct 2005, pp. 82–85

J. Berzowska, *Monograph. Seven Years of Design Research and Experimentation in Electronic Textiles and Reactive Garments* (XS Labs, Montréal, 2010)

B. Bilal, 3rdi (2013), http://wafaabilal.com/thirdi/. Accessed 05 Jan 2013

BMW (2008), GINA-The BMW group design philosophy. Challenging established concepts, Hazarding Visions. http://www.bmwusa.com/Standard/Content/AllBMWs/ConceptVehicles/GINA/. Accessed 05 Jan 2013

D. Bohde, Skin and the search for the interior: the representation of flaying in the art and anatomie of the Cinquecento, in *Bodily Extremities: Studies in Early European Cultural History*, ed. by F. Egmond, R. Zwijnenberg (Ashgate, Aldershot, 2003), pp. 10–48

M. Botvinick, J. Cohen, Rubber hands feel touch that eyes see. Nature **391**, 756 (1998)

F. Boussu, G. Bailleul, J.-L. Petitniot, H. Vinchon, Development of shape memory alloy fabrics for composite structures. AUTEX Res. J. **1**, 1–7 (2002)

S. Brady, B. Carson, D. O'Gorman, N. Moyna, D. Diamond, Body sensor network based on soft polymer sensors and wireless communications. J. Commun. **2**(5), 1–6 (2007)

C. Brower, R. Mallory, Z. Ohlman, *Experimental Eco-Design: Architecture, Fashion, Product* (RotoVision, Crans-Près-Céligny, Hove, 2005)

L. Bowie, Smart shirt headed to Smithsonian Institution. Whistle, Fac./Staff Newsp, Ga. Inst. Technol. **27**(19), 1 (2002)

L. Buechley, M. Eisenberg, J. Catchen, A. Crockett, The LilyPad Arduino: Using Computational Textiles to Investigate Engagement, Aesthetics, and Diversity in Computer Science Education. in *Proceedings of the SIGCHI Conference (CHI 2008)* April 2008. Florence, Italy, pp. 423–432 (2008)

U. Burke (2012), Una Burke website http://www.unaburke.com/. Accessed 05 Jan 2013

T. Carlyle, *Sartor Resartus: The Life and Opinions of Herr Teufelsdröckh* (Chapman and Hall, London, 1831)

M. Catrysse et al., Towards the integration of textile sensors in a wireless monitoring suit. Sens. Actuators, A **114**(2–3), 302–311 (2004)

G. Cho, J. Cho, C. Kim, J. Ha, Physiological and subjective evaluation of the rustling sounds of polyester warp knitted fabrics. Text. Res. J. **75**(4), 312–318 (2005)

A. Clark, *Natural-Born Cyborgs: Minds, Technologies, and the Future of Human Intelligence.* Oxford University Press (2003)

M.E. Clynes, N.S. Kline, Cyborgs and space. Astronautics. September 26–27 (1960)

M. Coelho, P. Maes, Sprout I/O: A texturally rich interface. in *2nd International Conference on Tangible and Embedded Interaction, Bonn*, 18–20 Feb 2008

S. Connor, *The Book of Skin* (Cornell University Press, Ithaca, 2003)

C. Darwin, *The Expression of the Emotions in Man and Animals* (John Murray, London, 1872)

G. Deleuze, *The Fold: Leibniz and the Baroque* (trans. by T. Conley). (Athlone Press, London, 1992)

M. DeMello, *Encyclopedia of Body Adornment* (Greenwood Press, Westport, 2007)

M. Douglas, *Natural Symbols* (Routledge, London, 2002)

P. Dourish, *Where the Action Is: The Foundation of Embodied Interaction* (The MIT Press, Cambridge, 2001)

G. Doy, *Drapery: Classicism and Barbarism in Visual Culture* (I.B Tauris, London, 2002)

B. Duchenne, *The Mechanism of Human Facial Expression or An Electro-Physiological Analysis of the Expression of the Emotions* (trans. by A. Cuthbertson). (Cambridge University Press, New York, 1990)

U. Eco, *Semiotics and the Philosophy of Language* (Indiana University Press, Bloomington, 1986)

C. Edwards, Wearable computing struggles for social acceptance; technology: The ultimate fashion item. IEE Rev. Oct 2003, 24–24 (2003)

J. Entwistle, *The Fashioned Body: Fashion, Dress and Modern Social Theory* (Polity, Cambridge, 2000)

J. Entwistle, The dressed body, in *Body Dressing (Dress, Body, Culture)*, ed. by J. Entwistle, E. Wilson (Berg, Oxford, 2001)

C. Evans, S. Menkes, T. Polhemus, B. Quinn, *Hussein Chalayan* (NAI Publishers/Groninger Museum, Rotterdam, 2005)

M. Featherstone, R. Burrows (eds.), *Cyberspace, Cyberbodies Cyberpunk: Cultures of Technical Embodiment* (Sage, London, 1995)

M. Fend, Bodily and pictorial surfaces: Skin in French art and medicine, 1790–1860. Art Hist. **28**, 311–339 (2005)

Fibretronic (2013), Fibretronic website. http://fibretronic.com/. 05 Jan 2013

E. Fiorani, *Abitare il corpo –il corpo di stoffa e la moda* (Lupetti, Milan, 2010)

J.C. Flugel, *Psychology of Clothes* (AMS Press, New York, 1930)

S. Freud, *The Ego and the Id* (trans. by J. Riviere). (Hogarth, London, 1927)

Y. Gao, Walking City (2012), http://yinggao.ca/eng/interactifs/walking-city/. Accessed 05 Jan 2013

W. Gaver, J. Bowers, A. Boucher, A. Law, S. Pennington, N. Villar, The History Tablecloth: Illuminating Domestic Activity. in *DIS 2006* (State College, Pennsylvania 2006)

W. Gibson, *Neuromancer* (ACE, New York, 1984)

E. Goffman, *The Presentation of Self in Everyday Life* (Anchor Books, New York, 1959)

R. Hahn, H. Reichl, Batteries and Power Supplies for Wearable and Ubiquitous Computing. in *Proceeding ISWC '99 3rd IEEE International Symposium on Wearable Computers 18–19 October 1999*, (San Francisco, California 1999)

D.J. Haraway, *When Species Meet* (University of Minnesota Press, Minneapolis, 2008)

R. Harper, T. Rodden, Y. Rogers, A. Sellen, *Being Human: HCI in 2020* (Microsoft, Cambridge, 2008)

A. Hollander, *Seeing Through Clothes* (University of California Press, Berkeley, 1993)

M.D. Husain, T. Dias, (2009), Development of Knitted Temperature Sensor (KTS). [pdf]. UK: University of Manchester,Textiles & Paper, School of Materials-SysTex student award 2009. http://www.systex.org/sites/default/files/SysTex_Student%20Award_Muhammad%20Dawood%20Husain_Report_0.pdf. Accessed 05 Jan 2013

D. Ihde, *Experimental Phenomenology* (Putnam, New York, 1977)

D. Ihde, A Phenomenology of Technics, in *Technology and Values: Essential Readings*, ed. by C. Hanks (Wiley-Blackwell, Chichester, 2010), pp. 135–155

iLuminate ™, iLuminate website (2012), http://iluminate.com/features/. Accessed 05 Jan 2013

H. Ishii, B. Ullmer, Tangible Bits: Towards Seamless. Interfaces between People, Bits and Atoms. in *Proceedings of. CHI '97, 22–27 March 1997* (ACM Press, Atlanta, Canada 1997)

J.E. Katz, Do machines become us?, in *Machines that Become Us: The Social Context of Personal Communication Technology*, ed. by J.E. Katz (Transaction Publishers, New Brunswick, 2003)

J. Kaur, C. Gale, *Fashion and Textiles: An Overview* (Berg, Oxford, 2004)

J.Y. Kim, (2012), Kinetic Mechanical Skirt. http://jaeyeopkim.com/Kinetic-Mechanical-Skirt. Accessed 05 Jan 2013

S. Klink, (2006), Philips Lumalive textiles light up the catwalk. http://www.newscenter.philips.com/main/standard/about/news/press/archive/2006/article-15518.wpd. Accessed 05 Jan 2013

M.I. Koo et al., Development of soft-actuator-based wearable tactile display. IEEE Trans. Rob. **24**(3), 549–559 (2008)

T. Kruger, Like a second skin: Living machines. Architect. Design **66**, 9–10 (1994)

K.L. LaBat, M.R. DeLong, Body cathexis and satisfaction with fit of apparel. Cloth. Text. Res. J. **8**(2), 43–48 (1990)

K. Lacasse, W. Baumann, *Textile Chemicals: Environmental Data and Facts (Engineering Online Library)* (Springer, New York, 2003)

R. Lane, B. Craig, Materials that sense and respond: An introduction to smart materials. AMPTIAC Q. **7**(2), 9–14 (2003)

S. Laverty, Hermeneutic phenomenology and phenomenology: A comparison of historical and methodological considerations. Int. J. Qual. Meth. **2**(3), (2003)

S. Lee, *Fashioning the Future: Tomorrow's Wardrobe* (Thames & Hudson, London, 2005)

M. Leenders, (2013), Marielle Leenders website. http://www.marielleleenders.nl/. Accessed 05 Jan 2013

J. Licklider, Man-Computer Symbiosis. in *IRE Transactions on Human Factors in Electronics, HFE-1*, pp. 4–11 (1960)

W. Lindeman, Y. Yanagida, H. Noma, K. Hosaka, Wearable vibrotactile systems for virtual contact and information display. Virtual Real **9**(2), 203–213 (2006)

Y. Liu, S. Gorgutsa, C. Santato, M. Skorobogatiy, Flexible, solid electrolyte-based lithium battery composed of LiFePO4 cathode and Li4Ti5O10 anode for applications in smart textiles. J. Electrochem. Soc. **159**(4), A349–A356 (2012)

G.O. Longo, Body and technology: Continuity or discontinuity?, in *Mediating the Human Body: Technology, Communication, and Fashion*, ed. by L. Fortunati, J.E. Katz, R. Riccini (Lawrence Erlbaum Associates, London, 2003), pp. 23–29

M. Lucy, H. Bart, (2008), Lucy and Bart Blogspot. http://lucyandbart.blogspot.com. Accessed 05 Jan 2013

Luminex® (2012), Luminex website. http://www.luminex.it/. Accessed 05 Jan 2013

E. Lupton, *Skin: Surface, Substance, and Design* (Princeton Architectural Press, New York, 2002)

M. Merleau-Ponty, *Phenomenology of Perception*. (trans. by C. Smith). (Routledge, London, 2005)

S. Mann, Smart clothing: The shift to wearable computing. Commun. ACM **39**(8), 23–24 (1996)

F.E. Mascia-Lees, P. Sharpe, *Tattoo, Torture, Mutilation, and Adornment: The Denaturalization of the Body in Culture and Text* (State University of New York Press, Albany, 1992)

M. McLuhan, *Understanding Media: The Extensions of Man* (McGraw Hill, New York, 1964)

M. McQuaid, *Extreme Textiles, Designing for high performances* (Thomas & Hudson, London, 2005)

D. Meoli, T. May-Plumlee, Interactive electronic textile development: A review of technologies. Spring **2**(2), 1–12 (2002)

Microsoft (2013), Xbox 360+Kinect. http://www.xbox.com/en-US/KINECT. Accessed 05 Jan 2013

H. Mieneke, Depicting skin: Microscopy and the visual articulation of skin interior 1820–1850, in *The Body Within: Art*, ed. by R. Vall, R. Zwijnenberg (Medicine and Visualization Brill, Leiden, 2009), pp. 51–66

I. Miyake, (2012), 132 5. ISSEY MIYAKE. http://www.isseymiyake.com/en/brands/132_5.html. Accessed 05 Jan 2013

M. Nagamachi, Kansei engineering: A new ergonomic consumer-oriented technology for product development. Int. J. Ind. Ergon. **15**, 3–11 (1995)

M. Nagamachi, Kansei engineering as a powerful consumer-oriented technology for product development. Appl. Ergon. **33**(3), 289–294 (2002)

Nendo (2006), Hanabi. http://www.nendo.jp/en/works/detail.php?y=2006&t=71. Accessed 05 Jan 2013

Nokia (2008), Nokia Research Center Nanotechnology: Big Potential in Tiny Substances. http://press.nokia.com/wp-content/uploads/mediaplugin/doc/nrc-nanotechnology-backgrounder.pdf. Accessed 05 Jan 2013

K. Ott, (2002) in *The sum of its parts: An introduction to modern histories of prosthetics* ed. by Ott, K, Serlin, D, Mihm, S. Artificial Parts, Practical Lives; Modern Histories of Prosthetics (New York University Press, New York, 2002)

K. Parsons, Merging Technology and Fashion. Stitch. Mag. (1999)

Peratech (2013), QTC^{TM} Material: Introduction to Quantum Tunnelling Composite. http://www.peratech.com/qtc-material.html. Accessed 05 Jan 2013

H. Perner-Wilson, M. Satomi, DIY Wearable technology. in *ISEA 2009* (Wearable Materialities Panel, Belfast UK, 2009)

Philips Design (2007), SKIN: Tattoo. http://www.design.philips.com/about/design/designportfolio/design_futures/tattoo.page. Accessed 05 Jan 2013

V. Pitts, *In the Flesh: The Cultural Politics of Body Modification* (Palgrave Macmillan, New York, 2003)

M. Polanyi, *Personal Knowledge-Towards a Post-Critical Philosophy.* (University of Chicago Press, Chicago, 1952)

T. Polhemus, *Hot Bodies, Cool Styles: New Techniques in Self Adornment* (Thames & Hudson, London, 2004)

T. Polhemus, L. Proctor, *Fashion & Anti-fashion: Anthropology of Clothing and Adornment* (Thames and Hudson, London, 1978)

Post R., Orth, M. (1997) Smart fabric or "Wearable Clothing". in *Digest of Papers, International Symposium on Wearable Computing* (IEEE Computer Society, Los Alamos, pp. 167–168)

B. Quinn, *Textile Futures: Fashion, Design and Technology* (Berg, Oxford, 2010)

S. Robertson, (2009), Transitional stripes. in *Proceedings of Smart Textiles Salon A Joint European Workshop*, 25 Sept 2009. Ghent, Belgium

D.L.T. Robles, Virtual Reality: Touch\Haptics, in *Sage Encyclopedia of Perception*, ed. by B. Goldstein (Sage Publication, Thousand Oaks, 2009)

M. Rodriguez, (2009), Skin-Transform. http://www.marisolrodriguezp.com/980x600/marisolrodriguez.html.Accessed 05 Jan 2013

M. Roustayi, Getting under the Skin: Rebecca Horn's sensibility machines. Arts Mag. May (1989)

S.E. Ryan, What is wearable technology art? Intell. Agent **8**, 1–6 (2008)

T. Schlömer, B. Poppinga, N. Henze, S. Boll, Gesture Recognition with a Wii Controller. in *International Conference on Tangible and Embedded Interaction (TEI 2008)*, 18–20 Feb 2008, Bonn, pp. 11–14 (2008)

S.J. Schwartzg, A. Pentland, The smart vest: Toward a next generation wearable computer platform. MIT Media Lab. Percept. Comput. Sect. Tech. Rep. **504**, 1–7 (1997)

S. Seymour, *Functional Aesthetics Visions in Fashionable Technology* (Springer, Vienna, 2010)

N. Shaari, M. Terauchi Kubo, H. Aoki, Recognizing female's sensibility in assessing traditional clothes. J. 6th Asian Design Int. Conf. **1**, 1348–7817 (2003)

B. Sherman, P. Judkins, *Glimpses of Heaven, Visions of Hell: Virtual Reality and Its Implications* (Hodder and Stoughton, London, 1992)

D.C. Simpson, in *The Choice of Control System for the Multimovement Prosthesis: Extended Physiological Proprioception (e.p.p.)*, ed. by P. Herberts, R. Kadefors, R. Magnusson, I. Petersén. The Control of Upper-Extremity Prostheses and Orthoses, Proceedings of the Conference on the Control of Upper-Extremity Prostheses and Orthoses, Springfield IL, (1974), pp. 146–150

Stelarc (1995), Extended-Body: Interview with Stelarc. Interviewed by Paolo Atzori and Kirk Woolford. Ctheory, 9 June 1995. http://www.ctheory.net/articles.aspx?id=71. Accessed 05 Jan 2013

H.Z. Tan, R. Gray, J.J Young, R. Traylor, A haptic back display for attentional and directional cueing. Haptics-e Electron. J. Haptics Res. **3**(1), 20 (2003)

X. Tao, Smart technology for textiles and clothing-introduction and overview, in *Smart Fibers, Fabrics and Clothing: Fundamentals and Applications*, ed. by X. Tao (Woodhead Publishing, Cambridge, 2001), pp. 1–6

X. Tao, Introduction, in *Wearable Electronics and Photonics*, ed. by X. Tao (Woodhead Publishing Ltd., Cambridge, 2005), pp. 1–12

M.C. Taylor, *Hiding (Religion and Postmodernism)* (University of Chicago Press, Chicago, 1997)

TCA (2004), Victimless Leather: A Prototype of Stitch-less Jacket grown in a Technoscientific 'Body'. http://tcaproject.org/projects/victimless/leather. Accessed 05 Jan 2013

Textronics (2013), Numetrex. http://www.numetrex.com/. Accessed 05 Jan 2013

E. Thorp, Optimal gambling systems for favorable game. Rev. Int. Stat. Inst. **37**(3), 273–293 (1969)

S. Turkle, *Life on the Screen: Identity in the Age of the Internet* (Simon & Schuster, New York, 1997)

B.S. Turner, *The Body and Society: Explorations in Social Theory* (Sage Publications, London, 2008)

D. Vatansever, E. Siores, R.L. Hadimani, T. Shah, in *Smart Woven Fabrics in Renewable Energy Generation,* ed. by S. Vassiliadis. Advances in Modern Woven Fabrics Technology Woven Fabrics. InTech. Ch. 7 (2011)

L. Verhoog, (2007), Fleshing Out, Living Fabrics for the Fashion Industry. A report by Leonieke Verhoog. http://www.v2.nl/archive/articles/fleshing-out/. Accessed 05 Jan 2013

X. Warm, (2012), WarmX-Functionality and technical principles. http://www.wellness.warmx.de /index.php/more-information.html. Accessed 05 Jan 2013

K. Warwick, M. Gasson, B. Hutt, I. Goodhew, P. Kyberd, B. Andrews, P. Teddy, A. Shad, The application of implant technology for cybernetic systems. Arch. Neurol. **60**(10), 1369–1373 (2003)

B. Wegenstein, *Getting Under the Skin: Body and Media Theory* (MIT Press, Cambridge, 2006)

M. Weiser, *The Computer of the Twenty First Century.* Scientific America, pp. 94–10 (1991)

M. Wigginton, J. Harris, *Intelligent Skins* (Butterworth-Heinemann, Oxford, 2002)

J.M. Woodham, *Twentieth-Century Design (Oxford History of Art)* (Oxford University Press, Oxford, 1997)

L. Worbin, Dynamic textile patterns, designing with smart textiles. Lic. Thesis. Department of Computer Science and Engineering, Chalmers University of Technology and The Swedish School of Textiles (2006)

J. York, P.C. Pendharkar, Human–computer interaction issues for mobile computing in a variable work context. Int. J. Hum-Comput. Stud. **60**, 771–797 (2004)

M. Zimniewska, I. Krucinska, The effect of raw material composition of clothes on selected physiological parameters of human organism. J. Text. Inst. **101**(2), 154–164 (2010)

Chapter 3
Emotion, Design and Technology

3.1 Perception of Emotion

Due to its Latin root that comes from *emovere*—to move away (Hillman 1960), emotion is not static but changes very quickly in time, while also makes human body to move, act and express. According to their duration and intention, Desmet (2002) divides affective states into four categories: emotions, moods, sentiments, and personality traits. Emotions and moods are short-lived affective states, while sentiments and personality traits are long-lived. Emotions have the shortest duration, as they can change in seconds spontaneously. These short-lived affective states can be classified according to their characteristics. Plutchik (1980) defined basic emotions as: acceptance, anger, anticipation, disgust, joy, fear, sadness, and surprise. Ekman (1999) synthesized basic emotions as: anger, disgust, fear, joy, sadness and surprise.

Besides defining what an emotion is, scientists have also tried to answer how emotions occur. While the central theories focus on the cognitive aspects of emotions, the peripheral theories stress on visceral level (Liss 1987). James-Lange theory proposed the idea that people experience emotions, because of the perception of visceral responses of their bodies (Denzin 1984). Cannon's (1927) theory disagrees with the James-Lange theory, and turned the attention into thalamus by saying that the emotion could occur without an arousal. Later, Schachter and Singer (1962) suggested that emotional experience requires both arousal and cognitive process. Damasio (1999) argues that emotions, feelings and consciousness are all related to human body and in the end they are felt in the somatic level through passing by a complex procedure that involves neural and chemical structures in the human body. While the human body and the mind are not separable but interconnected entities, emotions are strongly related to both corporeal and mental processes (Picard 1997). Emotion is lived through human body, where it is embodied as visceral sensation and form of expression. Human body is a medium of perception and a medium through which is perceived by others (Merleau-Ponty 2005). This dual function of human body plays an important role in experiencing an emotion. Picard (1997) divides the physiological

S. Uğur, *Wearing Embodied Emotions*, PoliMI SpringerBriefs,
DOI: 10.1007/978-88-470-5247-5_3, © The Author(s) 2013

changes caused by emotions as *apparent to others* (facial muscle movements, posture changes, etc.) and *less apparent to others* (heart rate, respiratory rate, etc.). All these aspects of emotions are intertwined during the perception of an emotion. The perception starts due to a trigger of an outside stimulus, a thought or a memory. Then, this stimulation passes through an intricate process that involves body arousal, behavioural response and appraisal.

3.1.1 Sensory Perception

Emotion can occur due to the perception of an outside stimulus. One of the important aspects of perception is the sensory process that is both kinetic and biochemical (Rodaway 1994). Sensory process starts where the stimuli come across the human body's sense organs, the gates of the mind. Each sense organ functions in a different manner and has varying dominances in perception of emotion. Although each sensory organ has a different role in sensory processing, they work cohesively in order to induce an emotion.

3.1.1.1 Touch

Touch is a basic human medium to perceive the outer world from the very close distance and a kind of emotional form of communication that plays major role in emotional development and social interaction. Due to the fact that touching requires an intimate interaction, this intimacy can alter the affection and trigger stronger emotions. There are two different tactile experiences: object perception via touch and social touch.

Object Perception via Touch
Touch is a complex sensorial experience that has various typologies, such as pressure, temperature, hardness, vibration or weight. These properties are perceived by two kinds of activities: passive and active touch. Passive touch can perceive temperature and pressure, while with active touch the hand actively moves through the object in order to recognize other properties, such as shape and texture (Gibson 1962). Haptic touch that is a combination of active and passive touch involves two kinds of receptors: the cutaneous receptors that are embedded in the skin and the kinaesthetic receptors that place in muscles, tendons, and joints (Lederman and Klatzky 2009). Cutaneous receptors are close to the surface of the skin and embrace mechanoreceptors that respond to mechanical stimulation (Loomis and Lederman 1986). Thermo-receptors respond to increases or decreases in skin temperature (Stevens 1991). On the other hand, kinaesthetic receptors are located in a deeper level of the skin and perceive the position and the movement of the body.

Lederman and Klatzky (1987) categorized the modalities of identifying an object by touch: *lateral motion* for texture, *pressure* for hardness, *static contact* for temperature, *unsupported hold* for weight, *enclosure* for volume of global shape, *counter following* for global shape, *part motion* for perception of moving parts and *function test* for discovering specific functions of the object. Object can have two properties that can be perceived through touch: macro-geometry, such as shape and size; and micro-geometry, such as surface, texture and hardness (Klatzky and Lederman 2003), and general perception occurs with the combination of the perception of these two properties.

Skin plays an important role in tactile perception. Different regions of the skin have different density of receptors (Cholewiak and Collins 2003), therefore touch sensitivity changes from area to area. Weber (1834/1978) conducted two-point discrimination test that measured the ability of various skin regions to perceive simultaneous two points of stimulation and found that fingertips and tips of tongue had more sensitivity. Weinstein (1968) also found that tactile spatial acuity is highest on the fingertips and lowest on the back.

Touch is an important sense for feeling of safety and pleasure (Ackerman 1990). Some objects can be perceived as pleasurable, some not, depending on their possible influence on the physical and psychological state. For instance, feeling pain or pleasure, directs a person to come closer to or stay distant from an object that they interact with. Ramachandran and Brang (2008) made studies on individuals that feel certain emotions when touching certain fabrics in order to display the link between tactile experience and emotions. Another study done by Gaëtan Gatian de Clérambault shows three cases of women patients, who displayed attraction for pieces of silk (Shera 2009). According to Sonneveld (2004), tactile experience can be divided into two characteristics: "intelligent and rational" experience that is the response to the physical qualities, such as shape, size, texture, weight, balance, temperature; and "dreamy and emotional" experience that creates emotional response due to the recall of a moment of previously lived emotion. An object can communicate a different meaning with its shape, but when it is touched, it can convey an emotion that recalls a tactile memory attached to that feeling of touch.

Social Touch
Touch not only functions for identifying an object; but also plays a major role in social interaction. From the very first stage of the human beings touch sense exists to experience and express emotions (Montagu 1986). Harlow's (1958) studies done with infant monkeys showed that the infants had more affectionate responses to a clothed, soft surrogate mother rather than to the one made from wires. For human infants affectionate touch is also crucial, hence the absence of it may cause psychological and physiological problems. When young children do not receive enough affectionate touch from their parents, they can show aggression during and later their childhood (Field 2001). The medical and psychological studies show that touching has an important affect in emotional development for not only infants

but also adults. For instance, touch can have positive influences in well being of the elderly (Bush 2001).

Touch is an intimate way of communication and; therefore, crucial for establishing and sustaining social bonds. As an immediate and intimate way of emotional expression it plays a crucial role in interpersonal relationships (Thayer 1986). Register and Henley (1992) found that it was easier to communicate intimate emotions with non-verbal tools such as touch, rather than with verbal tools. As language has words, human body has haptic codes that can express emotions in various modalities. Touch can convey hedonic feelings such as positive or negative emotions (Jones and Yarbrough 1985; Hertenstein et al. 2006; Knapp and Hall 1997). It can communicate positively warmth or negatively pain (Hertenstein et al. 2006). The perception of the social touch differs depending on the intimacy level of the relationship. Hertenstein et al. (2006) made a study on haptic codes with two groups of people, strangers and couples; and found that although for strangers it was difficult to distinguish touch that communicates love, sadness and sympathy; couples were less likely to confuse these emotions.

Touch has physical properties that vary in its action (e.g., rubbing, stroking, patting, pinching), intensity, velocity, abruptness, temperature, location, frequency, duration, and extent of surface area that is touched (Hertenstein 2002). Nguyen et al. (1976) found that playful and friendly touch was characterized with squeezing and patting, whereas stroking was associated with warmth or love and sexual desire. Hertenstein et al. (2006) found that sympathy was associated with stroking and patting; anger with hitting and squeezing; disgust with pushing; gratitude with shaking of the hand; fear with trembling, and love with stroking. Clynes (1977) by using *sentography*, which measures finger's vertical pressure and horizontal deflection in order to find the relation between the expressed emotion and the finger gesture (Picard 1997), found that while love was expressed with smooth pressure, hate and anger were expressed with particularly strong and instant pressure. Although these studies show that specific emotions are linked with specific haptic codes, cultural differences can affect the way that the haptic code is read. For instance, while two males hugging can be normal for a person from Europe, this could be found strange or not acceptable for a person from USA. Barnlund (1975) found that Japanese people would touch each other less than people in United States. On the other hand, not only culture, but also gender can determine the tactile communication limits. For instance, Heslin et al. (1983) showed that men found receiving touch from women strangers as pleasant, whereas women found touch that is applied by close friends of the opposite sex as pleasant. Another study indicates that males respond less positively to touch than women (Whitcher and Fisher 1979). Jones and Yarborough (1985) categorized body parts that receive touch: non-vulnerable body parts, such as hand, elbow, arm, shoulder and upper back; and vulnerable body parts, such as head, chest, genital area, and thighs. While hand, upper arms, forearms, shoulders, forehead and head are the areas, where friends mostly touch each other; women can feel sexual arousal from thighs, lips and chest (Jourard 1966).

Touch also works as a therapeutic element to cure psychological disorders or reduce stress. Affectionate touch, such as hug or massage can generate oxytocin that increases when stress related adrenalin hormones decrease (Light et al. 2005). For instance, regular and repeated warm touch between couples can reduce level of stress hormones (Holt-Lunstad et al. 2008). There are two kinds of touch according to the pressure deepness: *deep touch pressure* that includes types of firm touching, holding and stroking and *light touch pressure* that requires less pressure and is applied slightly, such as tickling. Occupational therapists have observed that very light touch alerts the nervous system, but deep pressure is relaxing and calming (Grandin 1992). For instance, heavy blankets and vests are used to prevent panic attacks and allow children and adults to resume normal activities (Champagne and Mullen 2005).

3.1.1.2 Vision

The recognition of emotions relies mostly on visual cues. Research on emotion recognition has been largely done through using visual tools, such as representations of facial expressions (Ekman and Friesen 1978, Ekman 1993). Visual perception helps people to give meaning and associations to artefacts and, therefore it is a crucial aspect for emotional appraisal. The appraisal consists of the judgment on *semantic interpretation* related to the way of use, *symbolic association* related to the meaning and *aesthetics* of the artefact that is related to beauty (Crilly et al. 2004).

Visual perception of an object and its affect on emotional appraisal rely on different visual properties: such as colour, size, or shape. Colour can evoke strong emotional responses. Levy (1984) claimed that warm colours could trigger arousal; on the other hand cold colours could reduce it. Jacobs and Hustmyer (1974) suggested that red was more arousing than green, green was more arousing than blue. Plutchik (1980) created a cone-shaped model (3D) and a wheel model (2D) to describe how emotions are related with colours and intensity. Mikellides (1990) claimed that chromatic strength could cause excitement or relaxation, rather than the color hue. Valdez and Mehrabian (1994) conducted a study on emotional reactions towards color hue, saturation, and brightness regarding to pleasantness, arousal and dominance. Boyatsiz and Varghese (1994) found that light colours (such as, blue and yellow) were associated with positive emotions and dark colours (such as, black and grey) were associated with negative ones.

Due to the fact emotions are communicated mostly with visual cues, such as facial and bodily expressions, shapes that refer to these gestural signs can be associated with emotions. For instance, Pavlova et al. (2005) found that instable shapes were associated with more fearful emotions and vertical shapes were mostly associated with joy. Larson et al. (2007) found that a simple "V" shape pointing downwards could be perceived as threatening. Aronoff et al. (1988) made a study on ceremonial masks in order to find which shape was associated threatening. They found that diagonal and angular elements were more threating than others and curved shapes could convey happiness (Aronoff et al. 1988). On the other hand, the study of Bar and Neta (2006) showed that sharp angles could create threating feeling, and therefore could be disliked. Leder and Carbon (2005) suggest that curved lines in car interiors can trigger positive emotions. Aesthetically pleasant things can attract people in a way that they can trigger positive emotions. From Greek philosophy to today, aesthetics has been discussed for centuries. Bell (1913–1961) claims that aesthetic qualities are the qualities in an object that evoke *aesthetic emotion*. Kawabata and Zeki (2004) found that beautiful pictures are correlated with activations in the areas of brain, which are associated with emotions. There is a strong connection between the visual aesthetics and emotions. According to Fiore and Kimle (1997), objects with low complexity are less pleasurable than objects with tolerable complexity. Cox and Cox (2002) also found that fashion apparel drawings that had moderate complexity were the most liked and with repeated visual exposures of these drawings complex designs were more likely to be trigger positive emotions than simple product designs. While a novel and complex shape can create uncertainty, repeated exposures can decrease this uncertainty and cause positive effects on perception of the visual stimulus (Berlyne 1970). Hekkert et al. (2003) found that people would prefer novel objects that didn't loose its prototypicality.

The natural way of expressing emotions is to move the body both intentionally and unintentionally, therefore movement can easily be perceived as a visual element that communicates emotions. According to Chafi et al. (2012) there are two kinds of motion that is related to emotions: biologically possible and biologically impossible. Human body is the basis for biological movement. Bacigalupi (1998) argues that objects that have human like kinetic properties can communicate emotions easier than abstract movements. Moving objects and animated artifacts are mostly found more exciting than the static ones. Study of Detenber et al. (1998) shows that moving clips can cause higher skin conductance and heart rate response rather than the still images. Pollick et al. (2001) found that the speed of the movement relies on the activation level. Lee et al. (2007) defined seven characteristics of movement: rhythm, beat, sequence, direction, path, volume and speed. According to framework of Tek-Jin et al. (2007), smooth, fast and open movements are suitable for expressing excited and happy emotions, besides disconnected and slow movements are suitable for expressing depressed and sad emotions.

3.1.1.3 Sound

Sound is one of the important stimuli that can elicit emotions. Baumgartner et al. (2006) found that when sound and vision are used together as emotional elicitors, they can create more influence on the emotional brain area rather than when they are singularly exposed. The characteristics of sound: loudness, roughness, and sharpness can result in decrease/increases of sensory pleasantness (Egmond 2008). Whether it is sensed through the ears or the eyes, rhythm is an important element of aesthetics. According to Meyer (1961), there are certain elements within the music, such as change of rhythm that create expectations about the future development of the music and, therefore, elicit arousal in the body. Music with fast tempo is generally associated with happiness and music with slow tempo is associated with sadness (Gagnon and Peretz 2003; Webster and Weir 2005).

On the other hand, rhythm perception can help to regulate the psychological state. Changes in pulse, respiration and heart rate can be induced by rhythm. Short scale rhythm is the time frame that corresponds to the human heart rate (Moravscik 2002). HeartMath's (2009) studies show that when people experience emotions, such as love, care, appreciation and compassion, the heart produces a more harmonious and balanced rhythm. Ries (1969) suggests that breathing frequency increases, when the person likes the music.

3.1.2 Body Arousal

The human body is the place, where the emotions are embodied as corporeal sensations. Each emotion affects the human body's metabolism in a different way. Autonomic Nervous System (ANS) that creates emotional arousal depending on the stimuli consists of sympathetic and parasympathetic systems. While sympathetic system prepares the body to cope with emotions, such as anger or fear, parasympathetic system calms the body down after a sympathetic activation (Kalat and Shiota 2006). Body arousal of an emotion can happen as changes in muscle tension, skin moister, skin temperature, blood flow, and heart rate or respiration rates. According to Ekman et al. (1983) negative emotions (anger, fear, sadness) could cause greater heart rate, than the positive emotions (happiness, surprise), and the finger skin temperature was evidently greater for anger, than the other emotions. Heart rate variability can give an idea about whether the person is in ease or not. As people experience feelings like anger, anxiety or fear, the heart rhythm patterns become more irregular. As sympathetic tone increases, the heartbeats get closer together, while parasympathetic tone increases, they widen out (Lipsenthal 2004). Gollnisch and Averill (1993) found that heart and respiratory rates increase during fear, anger and joy; while during sadness they decrease. Dishman et al. 2000 found that HRV is lower during stress. According to Cacioppo et al. 2000 diastolic blood pressure is greater during anger, than fear, sadness, and happiness.

On the other hand, GSR (Galvanic Skin Response) is a biological response that can increase with arousal. The more aroused a subject, the more the skin is moistened and, therefore conductivity increases. Palms and soles of the feet are the most sensitive places to detect skin conductance (Dawson et al. 2007). Gross and Levenson (1997) found that skin conductance increased after imagery of a video that elicits amusement and decreased after a neutral video.

3.1.3 Behavioral Response

People are social beings that are always in communication with each other through their bodies. While verbal-communication can deceive emotions (De Paulo et al. 2003), human body can give emotions out with a simple movement. Human body uses various ways to expose emotions. Muscle contractions on the face are most studied emotional expressions. Duchenne (1990), produced photographs of his subjects expressing distinct emotions by contracting their facial muscles. Just after him, Darwin (1872) studied facial muscles by using electric stimulation and reproduced the facial expressions in order to understand the expressions of basic emotions. Ekman and Friesen (1978) found Facial Action Coding System, which consists of taxonomies of emotional human facial expressions. Some facial expressions of emotion are universal across cultures (Frijda 1987).

Body gesture is also another medium for communicating emotions. Galton (1884) claimed that bodily inclination might be towards to a person, with whom there is a positive influence. Besides, De Rivera (1977) argues that there are four basic emotional movements: *toward other*, which is a positive inclination towards something; *toward self*, which is closer of the body positively; *away from self*, which is a negative inclination; and *away from the other*, which is a negative contraction. Pollick et al. (2001) analysed the relation between body movements and emotions and found that the speed of movement increases with an activated emotion. According to Coulson (2004) anger is predicted by bending backwards and arms forward; fear by head backwards and raising forearms; happiness by head backwards and raising arms above shoulder; sadness by bending forwards and arms at the sides; and surprise by bending backwards and raising arms with forearms straight. Cannon (1932) states that animals react to the threats with a general fighting or fleeing response. This behaviour can be seen in humans, as well. Sadness is mostly experienced as body without motion and closing off (Frijda 1987).

Also in language, people describe their emotion with movement metaphors; for instance being down for sadness, or up for being energetic and happy. Emotion can affect the way people do their actual actions, for instance how they walk or how strong they apply pressure. A heavy-footed walk can indicate anger or slower walk can communicate sadness (Evans-Martin 2007). Besides, Clynes (1977) with his study on *sentography* claimed that vertical and horizontal components of finger pressure could vary depending on which emotion is felt at the moment.

3.1.4 Appraisal

Besides the physiological experience, cognition plays an important role in perception of emotions. Parrot and Schulkin (1993) argue that emotions have a cognitive side that involves interpretation, memory and anticipation. Damasio (2006–1994) distinguishes emotions in two categories: primary emotions that are derived from direct stimuli resulting as physiological changes in the body, and secondary emotions that occur through various evaluations of the stimuli. Primary emotions, such as fear from sudden movements are instinctive and occur through thalamus and limbic system (Zimmermann et al. 2003). Besides, secondary emotions that occur in the cortex are learnt by time through the influence of many personal and social aspects, such as personality, culture, environment, etc. Appraisal theories claim that physiological and behavioral responses occur after the interpretation of the situation.

Cognition is the mental process involving remembering, association and recognition. People can react differently depending on their backgrounds, previous experiences, mental and physical states and many other personal and environmental factors. An appraisal can occur under the influence of the person's character depending on how he/she generally deals with the emotional situation. For instance, extraversion and neuroticism can effect the appraisal of emotions. Eysenck (1967) found that introverts and extraverts had different sensitivity levels in the cortical arousal system. Costa and McCrae (1980) found that the neuroticism strongly correlates with negative emotions and extraversion correlates strongly with positive emotions.

Cultural difference is also an important fact that influences the cognitive process of emotion. People from different countries might feel differently towards the same situations. Emotions can be learnt by the cultural values. *Cultural filter*, which consists of shared values, education level, age and socio-economic status (Jeans 1974), can affect the perception of the situation. For instance, Matsumoto (1992) found Japanese people (collectivist culture) showed lower accuracy for the display of negative emotions; on the other hand American people (individualistic culture) could tolerate the expression of negative emotions.

On the other hand, mood is another important factor that affects the cognitive process of emotions. While emotion is directed to a stimulus, mood is less intense than emotions and is often not directed to a certain situation or stimulus (Frijda 1993). The current mood state can influence the cognitive evaluation of the situation or the memory (Oatley and Jenkins 1996). For instance, Neumann et al. (2001) found that a person in happy mood could find the same stimulus funnier than a person in sad mood.

3.2 Functions of Expressing Emotions

Emotional expression is crucial for personal well-being and it can enhance social bonds by adding meaning to social interaction. Although there are four functions of emotions: individual (intrapersonal); dyadic (between two individuals); group (a set of individuals that directly interact); and cultural (within a large group that shares beliefs, norms, and cultural models) (Haidt and Keltner 1999), in this book the first two functions are addressed.

3.2.1 Intrapersonal Function

Expression of negative emotions plays an important role in intrapersonal level of emotional expression. If the negative emotions that are experienced as overwhelming for the receiver are not let off through expression, this can cause physical and psychological symptoms (Kennedy-Moore and Watson 1999). James et al. (1997) suggest that inhibiting negative emotions does not provide relief. Stress is one of the negative emotions, which has many deconstructive effects on the human body. Selye (1955) made a research on how stress affects the human body and categorized the steps of stress: *alarm* state, where the body gives alarm by producing adrenaline; *resistance* state, where the body resists to stress in order to cope with it; and *exhaustion* state, where the body cannot function normally. It is important to be aware of the body signals in the first stages of stress in order to avoid negative influences on the body.

3.2.2 Interpersonal Function

Emotional expression occurs mostly with the presence of others. For instance, infants produce their first emotional expressions, such as smiling, in a social interaction (Yale et al. 2003). Rimé et al. (1992) suggest that emotional experiences are mostly shared with others just after they occur in the body. People generally feel the need of sharing their emotions with other people. According to Izard et al. (1984) one of the social functions of expressing emotions is enhancing social interactions that can facilitate the development of interpersonal relationships.

There are some idioms for expressing emotions: such as letting ones feelings out or bottling them out. Bottling the emotions out is sometimes difficult or even not preferred in some certain situations and social norms. Hence, social context may affect the person's attitude of expressing emotions. It defers to express an emotion from being alone to being in public. Due to the fact that emotion is a social phenomenon, its expression rests on display rules and social motives. Emotional expression can occur with or without control. Pervasive belief claims that hiding emotions can indicate the strength or maturity (Kennedy-Moore and Watson 1999). For instance, in many cultures, a man crying cannot be acceptable. Fischer and Manstead (2000) argue that males in individualistic cultures use less emotional expressions, because expressing them might give negative impressions about their strength.

There are socially engaging and disengaging emotions. Planalp (1999) stated that when people experience strong emotions, such as joy and anger, they often feel urge to communicate it to others, on the other hand shame and regret cause opposite symptoms. Chapmann (1983) found that children, who watch cartoon alone, laughed less than the other children, who watched it with others. People can become more emotionally expressive or the opposite, when they know that other people are observing them. Moreover, emotional expressiveness depends on the level of intimacy or gender differences. Buck et al. (1992) found that the presence of a stranger in the environment can reduce the readability of emotional expressions, whereas the presence of a friend can increase. People express more intense emotions to the people, with whom they have an intimate relationship (Laurenceau et al. 1998), and women are more likely to show emotional expressions than men (Fischer and Manstead 2000).

Empathy refers to the ability of sharing the other person's emotions and feelings (Eslinger 1998). It is an important human behaviour that enhances interpersonal relationships and maintains social bonds. Mirroring is the behaviour, in which the person copies another person through mimicking the gestures, movements and body language during a conversations. Schachter (1959) claims that people can mimic the other person's emotional expressions. If someone shows a happy reaction towards a situation, this can influence the other person's expression to be positive.

3.3 Design and Emotion

Emotion plays a significant role in product design (Norman 2004; Jordan 2000; Desmet 2002). Today, costumers not any more buy just products, but experiences and emotions transmitted by these products (Jensen 1999). Today's designers shape intangible values and meanings in products and create personal experiences. Therefore, the user has become an important source for design process (McDonagh-Philip 1998). Design is not anymore just creating mere forms and thinking about efficiency but also constructing new stories for people to experience (Crossley 2003).

3.3.1 Emotions Derived by Products

Although an emotion is derived by social interaction or memory, a physical object can also trigger emotions due to its form, color, efficiency or behavior towards its user. Desmet (2002) introduced a basic "model of product emotions" that consists of three main elements: appraisal, stimuli and concern. Emotional appraisal occurs in order to interpret whether a stimuli is good for well-being or not (Lazarus 1991). In product design, appraisal is the interpretation of the product by its user. As a result of the appraisal, the product can be found as beneficial, harmful or not suitable to well-being (Desmet 2003). The appraisal can be done depending on the concerns of the user, such as the needs, goals and aims. Desmet (2003) defines five different product emotions: *instrumental* emotion that depends on whether the product lets the user achieve his/her goal or not; *aesthetic* emotion that is derived through appealing the senses; *social* emotion that depends on social norms and values; *surprise* that is related to the novelty of the product; and *interest* that attracts the user's attention. Besides, Rafaeli and Vilnai-Yavetz (2004) define three ways of sense making of the artifact: *instrumentality* that is related to the ease of use; *aesthetics* that refers to sensory reactions; and *symbolism* that refers to associations elicited by an artifact. Zajonc (1984) argues that primary emotions are reactions to aesthetics and secondary emotions are reactions to symbolic aspects of the product (Smith and Ellsworth 1985). Aesthetic experience, according to Desmet (2002), is related to product's capacity to appeal one or more of sensory modalities. For instance, an object can be visually beautiful, make a pleasant sound, have a pleasurable tactile quality, or smell nice. Besides, products can convey various symbolic values and meanings depending on their cognitive and social context (Krippendorff and Butter 1984). People can relate an object to a meaning through a cognitive process, such as metaphors, associations and past memories.

In 1980s many Italian companies have entered to a realm, where products appeal to the user's emotions with their forms that can recall past memories and associations. In furniture field designers, such as Geatano Pesce and Alessandro Mendini created extra-ordinary objects not only just to sit on, but also trigger affection, desire and attachment. On the other hand, Alessi that describes its self as the "Dream Factory" (Alessi 1998) has produced many products, which have their own characters and stories. With their human like colors, symbolic forms and particular materials the products have started to attract people by reaching their childhood memories in order to create playful experiences. Although this way of creating emotional products has had great success, there are more profound aspects of the emotions that are derived by products. The product should not just create a superficial fun on the user, but a more permanent experience through a well-designed interaction (Overbeeker et al. 2003).

Emotional experience is about the affective experience that the person obtains during the use of a product. Researchers have developed methodologies and measurement techniques in order to analyse product semantics in products design and what kind of emotions they trigger. Design researches generally rely on self-report methods (verbal and non-verbal questionnaires) for the measurement of user emotions. Desmet et al. (2000) introduce a non-verbal measurement tool that is called *PrEmo* in order to analyse emotions caused by product design through the use of expressive cartoon animations of dynamic facial, bodily, and vocal expressions. Kansei engineering is another approach for analyzing users emotions that are elicited by product design (Nagamachi 1995). Through these methodologies, users emotions towards products can be measured and this helps designers to get to know the user more deeply in order to design new products according to their emotional needs.

3.3.2 User Experience

Norman (2004) introduced three-levels of product experience: *visceral, behavioural* and *reflective. Visceral* level depends on the sensorial perception and is mostly related to the first impact of the appearance, touch and feel of the product. When a product is perceived, the stimuli first address the sensory organs. The sensory organs send the signals to the motor system in order to interpret whether the product is good for the well-being or not (Norman 2004). Behavioural level is about learned skills that are mostly subconscious (Norman 2004). It is related to the pleasure during the use. For instance, a product interface can create ease in use or can be problematic and complicated that exhausts its user. The more it is simple, the more it can communicate with its user. Reflective level is related to self-image and happens consciously (Norman 2004). Reflective process is mostly linked to the attachment of a product. For instance, a person can wear a watch from a specific brand that gives him\her additional status. Reflective process occurs in a social context.

People consciously or unconsciously choose which products make them happy. Pleasant emotions direct people to products that are pleasurable and beneficial, whereas unpleasant emotions push them from the products that are not good for their well-being (Desmet 2002). This is related to Frijda's theory (Frijda 1987) that claims people behave close or distant to an object according to their instinctive survival attitudes. Jordan (2000) has focused on pleasure and emotion in user experience. According to Jordan (2000) physio-pleasure is the physical pleasure that is derived from the sensorial experience of the object. It is related to sensorial experiences that are derived from material, colour, form, and texture of an object. Handheld objects have important tactile qualities that can alter physio-pleasure. Jacobs (1999) made a user test with an office desk in order to understand sensorial experience of the product and found that the users did not like the new product, because bad aroma of the glue caused disgust emotion. Sensorial characteristics of a product can have unexpected impacts that can create strong emotional affects on the user. Socio-pleasure is related to the pleasure, which is derived from social relationships (Jordan 2000). Some products can cause social interaction: for instance, a dress can attract attention of another person. Besides, there are products that are used by multiple users and cause social engagement. Psycho-pleasure is related to psychological state of the user (Jordan 2000). For instance, an object can create mental stress or relaxation while manipulating. The object can sometimes keep the psychological state of the user up, or sometimes makes the user feel disappointed or upset. Ideo-pleasure is linked to self-image and how the user is perceived by other people (Jordan 2000). For instance, the style or value of a product can affect the perception of its owner's image. According to Bourdieu (1991) there are two types of status: material status that expresses whether the user is wealthy or not and cultural status that expresses what kind of cultural norm the user belongs to. These two values can be conveyed by the products and alter the ideo-pleasure level.

Since interactive products have entered to the realm of design, engagement has also become an important factor in experience. A product should not only give aesthetic pleasure, but also involve the user more into action with engaging interactions (Hummels 1999). Today, while technology is increasingly penetrating into daily life, designer's role is becoming more and more complex. Dunne (1999) explains the designers' role as creators of alternative notions of use and need, rather than thinking about mere aesthetics and user friendliness. Four pleasure dimensions, which Jordan (2000) explains can be taken as an important basis in order to design today's technological products. Graver et al. (1999) constructed the idea of "cultural probes", which are tools for capturing human experience and reflecting aspects of human-product interaction in order to inspire the design concepts. New technologies have changed the traditional understanding of product design through adding interactive aspects that are embedded in the relation between product and user. "Technology probes" are design tools that are placed into the real context in order to collect data about how the user interact with technology and reflect the results as new ideas for human-technology interaction (Hutchinson et al. 2003).

3.4 Technology and Emotion

Although technology seems quite opposite to human emotions, today it can sense, interpret, learn emotional patterns and even act emotionally. While emotions have been largely addressed in product design field, this approach has been carried one step further with the influence of new technologies. Philips was one of the technology producers, which started to investigate research on emotional experience of electronic products (Marzano et al. 1995). The interplay between the product and the user is not any more static, but it has become a complex interaction that involves many different aspects, one of which is emotion. HCI (Human–Computer-Interaction) has shifted from operational dimension to experienced-based interaction, which involves emotions (Blythe et al. 2004). Emotion, pleasure and fun have become important research issues within HCI (Monk et al. 2002).

Emotional intelligence is the ability to recognize emotions of self and others, regulate these emotions and utilize them in order to give form to behaviour (Salovey et al. 2004). Emotional intelligence involves also empathy that is the ability to experience another person's emotion as one's own, and act accordingly. According to Goleman (1995) emotional intelligence is related to: *self-awareness* that is the ability to read someone's emotions, *self-management* that is the ability to control feelings, *social awareness* that is the ability to realize and react to others' emotions and *relationship managing* that is the ability to influence others during a conflict situation. In robotics there are many studies that work on emotional intelligence in robotic agents. *Kismet* is a robot that can interact with human beings through facial expressions (Breazeal and Aryananda 2002). Besides, *iCAT* can generate variety of facial expressions to simulate different emotions according to speech recognition (von Breemen 2005).

Today, emotional detection is possible via the use of variety of sensors. Technological agents can express emotions according to their users' emotions that are detected by these sensors. However, in human–machine interaction empathy—feeling the other person's emotion —is missing. Machines are not able to feel the same way that a person feels. Turkle (1997) made a research on people's opinion towards the idea of having a computer as a psychotherapist. Turkle (1997) found that the reporters would not rely on the computer, because computers could not understand the relationship between the human body and emotion. Emotions are felt and embodied in the body. Pain and pleasure are the visceral feelings that the emotion causes on the human body. It is questionable whether a computer feels pain or not. Dreyfus (1967) questioned if computers would need bodies in order to be intelligent. According to Picard (1997) computers have different bodies that are constructed from cameras, microphones, keyboards and sensors, rather than having skins and bones. Turkle (1997) argues that human body is different than computers, because it can "feel" but computers cannot. Today, computers have gained emotional intelligence, and moreover, with the improvements in tissue engineering they can have human like communicative bodies. But a question appears: Do people rely on them? According to Picard (1997), computers do not need to become humanoids to act emotionally. In 1919, Freud (2003) mentioned about "uncanny" concept that means being familiar and at the same time foreign to a living organism. Mori (1970) carried this concept into robotics with his theory of *the uncanny valley* that indicates the negative emotional response towards robots that seem and act like human. When the human likeness and movement of the robot is more than it should be, then this can cause negative effects on emotional response.

McLuhan (1964), who refers technology as social and psychic "extensions" of the human body mentions the story of *Narcissus* from mythology as an example of explaining people's interest on technology. People as *Narcissus* can be fascinated with their extensions, things that look like them. Lacan theory on *mirror stage* (Lacan 1977)—the first step for self-identity development—can also explain people's reaction to new technologies that are similar to their nature (Webster 1994). Today, with the new technologies *mirror stage* happens, when people face with the technologies that are similar to their nature, as if they face themselves for the first time. Therefore, this can create a kind of fascination towards technology. However, the designer of new technologies should bear in mind that mimicking the human body's physical appearance might push them into the uncanny valley and, therefore create negative emotional responses. Using metaphors in design can be useful, but at the same time can cause backlash (Norman 2007). The product should have its own identity, rather it resembles to some other artefact through metaphors (Blythe et al. 2004). Technology can turn the world into a "techno-logical zoo", but on the other hand can make it a beautiful place to live (Dunne 1999). Between these two opposite directions, the role of the designer is to create artefacts that are not only coherent to the human body, but also push the limits of stereotypes by inventing new forms and interaction scenarios.

3.4.1 Measuring Emotions Via Technology

Affective computing is an interdisciplinary field, which focuses on computers that can measure, influence and regulate emotions (Picard 1997). It can be utilized in various areas, such as marketing to measure customer interest, health care to assist psychological disorders and communication to create ease for expressing emotions. Emotion can occur due to a complex integration of various parameters, such as cognitive, physiological or environmental. There is a wide range of studies on measuring emotions in varying modalities, such as facial expression, speech, gesture or biological data detection. For more reliable measurement, different methods can be merged. For instance, a person can have an increasing heart rate but it cannot point out the happiness by its own. Other parameters are needed to measure the exact emotion.

Russell (1980) introduced the circumplex model of affect that consists of two fundamental dimensions: valence (pleasure–displeasure) and arousal (passive–active). Emotion, the cause of pain or pleasure (Schachter and Singer 1962), has a hedonic dimension, ranging between positive and negative values as valence parameters. On the other hand, arousal is related to the body response towards an emotion. It ranges from calming to exciting, or passive to active. The autonomic nervous system increases the heart rate and blood pressure with an aroused emotion. According to Schachter and Singer (1962) arousal can happen without emotion, but emotion cannot be emerged without arousal. Many studies based their experiments on arousal-valance diagrams. In many studies, valance level can be detected by heart rate, while arousal level can be detected by skin conductance (Peter 2008). Due to the fact that irregular respiratory pattern is caused by negative emotions; it can be used as indicator of valance (Kim and André 2008). GSR device can measure skin conductance that increases with a high sympathetic activity and infrared emitter-receiver finger clips can measure the Blood Volume Pulse (BVP) referring to Heart Rate Variability (HRV) (Verhoef et al. 2009). Electromyography (EMG) that measures muscle activity can be used as an indicator of stress (Melin and Lundberg 1997).

Although body's vital signals, such as skin conductance, heart rate, blood pressure, and respiration can provide information regarding to the emotional state, each person has different physiological system and this can cause difficulties on measuring emotions based on specific parameters. For instance, one can have different heart rate variability for stress than another person because of the metabolism, gender or age differences. Due to the fact that biological measurement cannot always give accurate results, a combination of measurement techniques should be applied in order to obtain realistic results. Although there are many studies based on dimensional theories, appraisal theorists argue that in order to distinguish emotions, the interpretation of the person is crucial (Scherer 2005). Morris (2011) suggests that cognitive method -self-reporting- could be merged with physiological data that is tracked by biosensors.

Not only bio-signals, but also behavioural signs of the human body can be tracked in order to measure emotions. For instance, facial expression detection is one the most used techniques to measure emotions. Facial EMG technique is a tool for measuring facial muscle activities during emotional expressions. This technique is used to measure the electrical activity of the facial muscles that is created by muscle fibres in contraction. The valance of emotion can be detected by tracking the contraction of facial muscles -zygomatic and corrugator (Peter 2008). On the other hand, speech detection is one of the techniques that deal with emotional measurement. While fear and anger can be expressed as fast and laud speech; emotions, such as tiredness, boredom or sadness cause to slower and lower-pitched speech (Breazeal and Aryananda 2002). Jones and Jonsson (2008) created a system, which detects the speech of the driver in order to analyse the emotions during the drive and responds to the driver in a more intuitive way. Gestures, as well, can be detected and analysed in order to measure the emotional state of the person. In their e-learning study, Mota and Picard (2003) measured emotions by posture information derived from sensors that are attached in a chair. Zimmermann et al. (2003) developed a measurement method that was done through keyboard and mouse that detected the motor-behavioural parameters. *Sentic mouse* that is a computer mouse with a force resistor detects dynamic finger pressure in order to capture valence value (Krish 1997).

With the possibility of wearable wireless body networks, emotions might be detected and communicated in order to enhance the wearer's psychological state and create new communication patterns in interpersonal relationships. Although non-contact techniques can be done in order to monitor body movements or facial expressions, compare to wearable sensors they can track less information of vital signals, such as heart or respiration rates (Brady et al. 2007). Due to the fact that wearable wireless body sensors let the wearer act freely, the detection of body signals becomes easier and less disturbing. Wearable monitoring can be done by textile-based sensors embedded into garments that are worn as tight suits or local wearable devices, such as bracelets or gloves. Bio-sensing technologies have opened new application areas and turned the classical healthcare system into a more flexible and home-based state. Body sensor healthcare applications can monitor vital signs, provide feedback to the user, and send this data to a medical assistant. In wireless body sensor network systems the communication is done via mobile phones, PDA devices, laptops and PCs that are connected through wireless personal network (ZigBee or Bluetooth). This technology can also be used to track emotional state of the person and even store this data in a server that analyses the changes in psychological state over time.

In MIT, Picard and her team worked on wireless wearable solutions for affect detection. They developed *Affectiva-Q Sensor* (Affectiva Inc 2012) that can sense skin conductance, motion and environmental temperature in order to measure wearer's emotional state. Motion and environmental temperature sensors are used for avoiding wrong measurement of arousal. This bracelet shape sensor has shifted the idea of emotional measurement in the laboratory settings into everyday life environments. For mobile tracking of emotions, Peter et al. (2005) developed a wearable system that consists of heart rate sensor worn on the chest and skin temperature and skin conductivity sensors integrated into a glove. On the other hand, The *Affective Diary* collects affective data of the wearer through a biosensor armband and visualizes this data on an interface, where the emotions are shown as abstract colorful forms (Ståhl et al. 2008).

3.4.2 Technology Mediated Emotional Communication

Mobile communication technologies offer the possibility of connecting with others in anywhere and anytime. This ability can also disconnect their physical-self from their immediate surroundings. "Tele-cocooning" is a term that was used by Habuchi (2005) to describe intimate human-computer interaction, which refers to communication without physical interaction. The concept of "tele-cocooning" can be seen everywhere in daily life as an enclosing relationship with technology. Furthermore, Internet has been reshaping the social bonds and dynamics. Due to the shift of face-to-face communication to digitally mediated communication, the bonds in network society are becoming looser compared to traditional communities (van Dijk 1997). Digital communication technologies and Internet not only affect the social relationships, such as friendship circles and family relationships, but also blurs the certain boundaries between public and private (Katz 2006). While the social structures are getting flexible and fuzzy, people may have the need to delink themselves in their close relationships, which are mostly built on intimacy and trust that is generally obtained through the exercise of the emotional intelligence (Vincent and Fortunati 2009).

Picard (1997) describes the "affective bandwidth" as how much affective information passes through the communication channel. According to Picard (1997), email communicates the least affect than phone, than videoconference, and the most affect is mediated in face-to-face communication. A "mediated emotion" is an emotion that is produced in a telephone or mobile phone conversation via a computational electronic device (Vincent and Fortunati 2009). Due to the lack of facial expression, body language, tone of voice, clothing, physical surroundings, and other contextual information, Net Generation has to create solutions in order to express their emotions within the limitations of their keyboards (Tapscott 1997). People broadly use the ASCII-based codes known as emoticons in chat platforms (O'Hagan and Ashworth 2002). According to Tsetserukou and Neviarouskaya (2010), a conventional emotional mediation usually lacks visual emotional signals, such as facial expressions, gestures and ignores tactile social communication. On the contrary, Derks et al. (2008) argues that when visibility of emotional expression is limited, this can create ease in communicating emotions compare to face-to-face communication. People can feel more relaxed and frank about their emotions, when they are communicating via digital agents.

Computer-mediated communication supported by affective computing can introduce new ways of expressing emotions. *EmoteMail* that captures photos of facial expressions uses color-codes on the text in order to convey the writer's emotions to the reader (Angesleva et al. 2004). Embodied Conversational Agents (ECA) research concentrates on developing computer-generated characters that are able to do facial expressions and body gestures that convey emotions (Lee and Marsella 2006). On the other hand, Pantic et al. (2007) envisions that interaction can widen the limits of the physical interface of the computing technologies and shift to a more human like state through measuring and responding to behavioural and social signalling. New communication technologies can open new avenues to express emotions and change the traditional social norms and habits. Besides, due to its intimate interaction with human body, wearable technology can be used in mediated communication in order to detect, express and regulate emotions.

3.4.2.1 Dynamic Wearable Interfaces

Emotion is always related to motion. Body motion can be easily measured by wearable interfaces due to the fact that they are placed on the body. KUIs (Kinetic User Interfaces) can capture user's motion and behavior in order to create an engaging interaction (Pallotta 2009). KUIs can be utilized in communication technologies in order to track emotional expressions via body gestures and translate this information to an emotional message.

On the other hand, a dynamic artefact that can move and transform into new shapes and colours can convey emotions. Microelectronics, mechanics, shape-changing materials and OLED displays can be embedded into soft and flexible surfaces in order to create organic interfaces that respond to any input modality. Organic user interfaces (OUIs) combines the input and output in one device that is activated by touch, gestures or deformation (Holman and Vertegaal 2008). The difference of organic tangible interfaces from tangible interfaces is that the display can change shape and state during the interaction. These interfaces can carry the traditional HCI into a dynamic level, where the interface behaves as a living entity that can convey meaning and messages through addressing majority of the sensory organs. According to Parkes et al. (2008), Kinetic Organic Interfaces (KOIs) are organic user interfaces that give outputs as kinetic motion in order to communicate the input. These interfaces can easily elicit emotional responses in user. Dynamic interfaces that are worn on the body can be used in order to convey immediate emotional messages. This type of interfaces can be mediums to communicate emotions to the nearby people. By turning the body into a dynamic display, these interfaces can change the way, in which emotions are normally expressed. The *Chimerical Garment* translates biological information, such as body temperature, breathing and movement into abstract visual signals that is worn on the body via embedding LCD on the garment (Dee Co 2000). Stead et al. (2004) developed

The Emotional Wardrobe that is an integration of affective computing and fashion in order to represent emotional response on the surface of clothing as flexible visual displays. On the other hand, Philips (2006) developed *SKIN: Dresses* that can change pattern and color with its light-emitting surface in order to convey emotional state of the wearer depending on measurements of the biosensors. *Exhale* is a wearable body network project that conveys the affective data through the small fans, speakers and vibrators beneath the skirts of the wearers according to wearers' breathing patterns (Schiphorst 2005).

3.4.2.2 Affective Wearable Haptics

Haptic interfaces can be used in order to induce and communicate emotions in mediated communication. Due to the fact that touch is an intimate form of emotional expression, affective haptics can strength relationships and add more intimacy. Affective haptics is an emerging area of research, which focuses on design of devices and systems that can "elicit, enhance, or influence" the emotional state of the user through touch (Tsetserukou and Neviarouskaya 2010). Mediated social touch is the ability to touch another person over a distance by means of tactile or kinesthetic feedback technology (Haans and Ijsselsteijn 2006). There are different types of tactile simulation devices, such as electromagnetic displays, pneumatic displays, displays with SMA, piezo-electronic displays or other actuator mechanisms (Allerkamp 2010). For communicating emotions through haptic devices, Rovers and van Essen (2004) developed a framework that uses *hapticons*—small force patterns. Chang et al. (2002) developed *ComTouch* that can transmit finger pressure as vibro-tactile stimuli over a distance. *LoveBomb* is another device that has different vibro-tactile stimuli to represent love and sadness (Hansson and Skog 2001).

Wearable technology's intimate interaction with human skin leads to new design solutions where the tactile stimuli are directed to the human body from a very close distance. Wearable technology with haptic features can introduce new avenues for communicating emotions over distance. For instance, Hug Shirt™ was designed in order to transmit the sensation of touch over a distance by wearable agents that have sensors for measuring the strength of the touch, skin temperature, heartbeat rate of the wearer and transmitters of touch and warmth (cutecircuit 2013). *TapTap* is a wearable haptic interface that can record and play back patterns of touch in order to experience the affective human touch (Bonanni et al. 2006). *iFeel_IM!* (Intelligent system for Feeling enhancement powered by affect sensitive Instant Messenger) is a wearable system that reproduces various types of social touch in 3D virtual world in order to enhance user's affective state (Tsetserukou and Neviarouskaya 2010).

References

D. Ackerman, *A Natural History of the Senses* (Vintage Books, London, 1990)

Affectiva Inc (2012) Affectiva QTM solutions white paper-liberate yourself from the Lab: Q Sensor measures EDA in the wild. [pdf] http://www-assets.affectiva.com/assets/Whitepaper_Q_Measuring_EDA_in_the_Wild.pdf. Accessed 5 Jan 2013

A. Alessi, *The Dream Factory: Alessi Since 1921* (Konemann, Köln, 1998)

D. Allerkamp, *Tactile Perception of Textiles in a Virtual-Reality System* (Springer, Berlin, 2010)

J. Angesleva, C. Reynolds, S. O'Modhrain, EmoteMail. in *Proceedings of SIGGRAPH 8–12 Aug 2004* (ACM, Los Angeles 2004)

J. Aronoff, A.M. Barclay, L.A. Stevenson, The recognition of threatening facial stimuli. J. Pers. Soc. Psychol. **54**, 647–655 (1988)

M. Bacigalupi, The craft of movement in interaction design. in *Proceedings of AVI98 Advanced Visual Interface Conference 25–27 May 1998*. L'Aquila (1998)

M. Bar, M. Neta, Humans prefer curved visual objects. Psychol. Sci. **17**, 645–648 (2006)

D.J. Barnlund, Communication Styles in Two Cultures: Japan and the United States, in ed. by A. Kendon, M.R. Harris, M.R. Key, *Organization of Behavior in Face-to-Face Interaction.* (Mouton, The Hague, (1975)

T. Baumgartner, M. Esslen, L. Jancke, From emotion perception to emotion experience: emotions evoked by pictures and classical music. Int. J. Psychophysiol. **60**, 34–43 (2006)

C. Bell, Art. (Arrow Books, London 1913–1961)

D.E. Berlyne, Novelty, complexity and hedonic value. Percept. Psychophys. **32**, 279–286 (1970)

M.A. Blythe, K. Overbeeke, A.F. Monk, P.C. Wright, *Funology: From Usability to Enjoyment* (Kluwer Academic Publisher, Dordrecht, 2004)

L. Bonanni, J. Lieberman, C. Vaucelle, O. Zuckerman, TapTap: A Haptic Wearable for Asynchronous Distributed Touch Therapy, in *CHI 2006, 22–27 Apr 2006* Montréal (2006)

P. Bourdieu, *Language and Symbolic Power,* ed. by JB. Thompson (Trans. by G. Raymond, M. Adamson) (Harvard University Press, Cambridge, 1991)

C.J. Boyatzis, R. Varghese, Children's emotional associations with colors. J. Genet. Psychol. **155**, 77–85 (1994)

S. Brady, B. Carson, D. O'Gorman, N. Moyna, D. Diamond, Body sensor network based on soft polymer sensors and wireless communications. J. Commun. **2**(5), 1–6 (2007)

C.L. Breazeal, L. Aryananda, Recognition of affective communicative intent in robot-directed speech. Auton. Robots **12**(1), 83–104 (2002)

R. Buck, J.I. Losow, M.M. Murphy, P. Costanzo, Social facilitation and inhibition of emotional expression and communication. J. Pers. Soc. Psychol. **63**(6), 962–968 (1992)

E. Bush, The use of human touch to improve the well-being of older adults. J. Holist. Nurs. **19**, 256–270 (2001)

J.T. Cacioppo, G.G. Berntson, J.T. Larsen, K.M. Poehlmann, T.A. Ito, The psychophysiology of emotion, in *Handbook of Emotions*, 2nd edn., ed. by M. Lewis, J.M. Haviland-Jones (Guilford Press, New York, 2000), pp. 173–191

W.B. Cannon, The James-Lange theory of emotion: a critical examination and an alternative theory. Am. J. Psychol. **39**, 10–124 (1927)

W.B. Cannon, *The Wisdom of the Body* (Norton Pubs, New York, 1932)

A. Chafi, L. Schiaratura, S. Rusinek, Three patterns of motion which change the perception of emotional faces. Psychology **3**, 82–89 (2012)

T. Champagne, B. Mullen, The Weighted Blanket: Use and Research in Psychiatry, in: *MAOT'05* (2005)

A. Chang, S. O'Modhrain, R. Jacob, E. Gunther, I. Hiroshi, ComTouch: Design of a vibrotactile communication device, *in Proceedings of DIS.* (ACM Press, New York 2002)

A.J. Chapmann, Humor and laughter in social interaction some implications for humor research, in ed. by P.E. McGhee, J.H. Goldstein, *Handbook of Humor Research.* (Springer, New York, 1983), pp. 35–157

R.W. Cholewiak, A.A. Collins, Vibrotactile localization on the arm: effects of Place, Space, and Age. Percep. Psychophys. **65**(7), 1058–1077 (2003)

M. Clynes, *Sentics: The Touch of Emotion* (Anchor Press/Doubleday, Garden City, 1977)

P.T.J. Costa, R.R. McCrae, Influence of extraversion and neuroticism on subjective well-being: Happy and unhappy people. J. Pers. Soc. Psychol. **38**, 668–678 (1980)

M. Coulson, Attributing emotion to static body postures: recognition, accuracy, confusions and viewpoint dependence. J. Nonverbal Behav. **28**(2), 117–139 (2004)

D. Cox, A.D. Cox, Beyond first impressions: The effects of repeated exposure on consumer liking of visually complex and simple product designs. J. Acad. Mark. Sci. **30**(2), 119–130 (2002)

N. Crilly, J. Moultrie, P.J. Clarkson, Seeing things: consumer response to the visual domain in product design. Des. Stud. **25**(6), 547–577 (2004)

L. Crossley, Building emotions in design. Des. J. **6**(3), 35–45 (2003)

Cutecircuit (2013) Hug Shirt http://www.cutecircuit.com/hug-shirt/. Accessed 5 Jan 2013

A. Damasio, *The Feeling of What Happens: Body and Emotion in the Making of Consciousness* (Harcourt Brace, New York, 1999)

A. Damasio, *Descartes' Error: Emotion, Reason and the Human Brain*. (Vintage Books, London, 2006/1994)

C. Darwin, *The Expression of the Emotions in Man and Animals* (John Murray, London, 1872)

M.E. Dawson, A.M. Schell, D.L. Filion (2007) The electrodermal system, in *Handbook of Psychophysiology*, ed. by J.T. Cacioppo, L.G. Tassinary, G.G. Berntson (Cambridge University Press, Cambridge) pp. 159–181

J.H. De Rivera, A structural theory of the emotions. Psychol. Issues Monogr. **40**, 169 (1977)

E. Dee Co, *Computation and Technology as Expressive Elements in Fashion. Master Thesis* (MIT: Media Arts and Sciences, Cambridge, 2000)

N.K. Denzin, *On Understanding Emotion* (Jossey-Bass Publishers, San Francisco, 1984)

B.M. DePaulo et al., Cues to deception. Psychol. Bull. **129**(1), 74–118 (2003)

D. Derks, A.H. Fischer, A.E.R. Bos, The role of emotion in computer-mediated communication: a review. Comput. Hum. Behav. **24**, 766–785 (2008)

P.M.A. Desmet (2002) Designing Emotions. Ph.D. Thesis, Delft University of Technology

P.M.A. Desmet, A multilayered model of product emotions. Des. J. **6**(2), 4–11 (2003)

P.M.A. Desmet, P. Hekkert, J.J. Jacobs, When a car makes you smile: Development and application of an instrument to measure product emotions. Adv. Consum. Res. **27**, 111–117 (2000)

B.H. Detenber, R.F. Simons, G.G. Bennett, Roll 'em! The effects of picture motion on emotional responses. J. Broadcast. Electron. Media **42**, 113–127 (1998)

R.K. Dishman, Y. Nakamura, M.E. Garcia, R.W. Thompson, A.L. Dunn, S.N. Blair, Heart rate variability, trait anxiety, and perceived stress among physically fit men and women. Int. J. Psychophysiol. **37**, 121–133 (2000)

H. Dreyfus, Why computers must have bodies in order to be intelligent. Rev. Metaphys. **21**, 13–32 (1967)

B. Duchenne, *The Mechanism of Human Facial Expression or an Electro-Physiological Analysis of the Expression of the Emotions* (Trans. by A. Cuthbertson). (Cambridge University Press, New York 1990)

A. Dunne, *Hertzian Tales: Electronic Products, Aesthetics Experience and Critical Design* (RCA CRD Research Publications, London, 1999)

R.V. Egmond, Impact of sound on image-evoked emotions, in ed. by B.E. Rogowitz, T.N. Pappas Human Vision and Electronic Imaging XIII Vol. 6806. *Proceedings of SPIE- International Society for Optical Engineering*. (SPIE, Bellingham, 2008) pp. 1–12

P. Ekman, Facial expression and emotion. Am. Psychol. **48**, 384–392 (1993)

P. Ekman, Basic Emotions, in ed. by T. Dalgleish, T. Power, *The Handbook of Cognition and Emotion*. (Wiley, Sussex 1999), pp. 45–60

P. Ekman, W.V. Friesen, *Facial Action Coding System* (Consulting Psychologists Press, Paolo Alto, 1978)

P. Ekman, R.W. Levenson, W.V. Friesen, Autonomic nervous system activity distinguishes among emotions. Science **221**(4616): 1208–1210 (1983)

P.J. Eslinger, Neurological and neuropsychological bases of empathy. Eur. Neurol. **39**(4), 193–199 (1998)

F.F. Evans-Martin, *Emotion and Stress* (Chelsea House, New York, 2007)

H.J. Eysenck, *The Biological Basis of Personality* (Thomas Springfield, Llinois, 1967)

T. Field, *Touch Therapy* Harcourt Brace, New York (2001)

A.M. Fiore, P.A. Kimle, *Understanding Aesthetics for the Merchandising and Design Professional* (Fairchild Publications, New York, 1997)

A.H. Fischer, S.R. Manstead, The relation between gender and emotions in different cultures, in ed. by A.H. Fischer, *Gender and Emotion: Social Psychological Perspectives.* (Cambridge University Press, Paris, 2000) pp. 71–94

S. Freud, *The Uncanny* (Translated by McLintock D, Penguin Classics, London, 2003)

N.H. Frijda (1987) *The Emotions (Studies in Emotion and Social Interaction).* Cambridge University Press, Cambridge

N.H. Frijda, Moods, Emotion Episodes and Emotions, in ed. by M. Lewis, J.M. *Haviland, Handbook of Emotions.* (Guilford Press, New York, 1993) pp. 381–403

L. Gagnon, I. Peretz, Mode and tempo relative contributions to "happy-sad" judgements in equitone mequitone. Cogn. Emot. **17**, 25–40 (2003)

F. Galton, Measurement of character. Fortn. Rev. **42**, 179–185 (1884)

J.J. Gibson, Observations on active touch. Psychol. Rev. **69**, 477–491 (1962)

D. Goleman, *Emotional Intelligence: Why It Can Matter More than IQ* (Bantam, New York, 1995)

G. Gollnisch, J.R. Averill, Emotional imagery: Strategies and correlates. Cogn. Emot. **7**, 407–429 (1993)

T. Grandin, Calming effects of deep touch pressure in patients with autistic disorder, college students, and animals. J. Child Adolesc. psychopharmacol. **2**, 63–70 (1992)

B. Graver, A. Dunne, E. Pacenti, Cultural Probes. Interaction **6**(1), 21–29 (1999)

J.J. Gross, R.W. Levenson, Hiding feelings: the acute effects of inhibiting negative and positive emotions. J. Abnorm. Psychol. **106**(1), 95–103 (1997)

A. Haans, W. Ijsselsteijn, Mediated social touch: a review of current research and future directions. Virtual Real. **9**, 149–159 (2006)

I. Habuchi, *Accelerating Reflexivity, in Personal* (Mobile Phones in Japanese Life. Edited by Ito M, Okabe D, Matsuda M. MIT Press, Cambridge MA, Portable, Pedestrian, 2005)

J. Haidt, D. Keltner, Social functions of emotions at four levels of analysis. Cogn. Emot. **13**(5), 505–521 (1999)

R. Hansson, T. Skog, The LoveBomb: Encouraging the communication of emotions in public spaces, in *Computer-Human Interaction (CHI)*, Extended Abstracts. ACM Press, Seattle (2001)

H.F. Harlow, The nature of love. Am. Psychol. **13**, 673–685 (1958)

HearthMath LLC (2009) An Appreciative Heart is Good Medicine. http://www.heartmath.com/articles/appreciative-heart-good-medicine.html. Accessed 5 Jan 2013

P. Hekkert, D. Snelders, P. Van Wieringen, Most advanced, yet acceptable: Typicality and novelty as joint predictors of aesthetic preference in industrial design. Br. J. Psychol. **94**(1), 111–124 (2003)

J.M. Hertenstein, Touch: its communicative functions in infancy. Hum. Dev. **45**, 70–94 (2002)

J.M. Hertenstein, J.M. Verkamp, A.M. Kerestes, R.M. Holmes, The communicative functions of touch in humans, nonhuman primates, and rats: a review and synthesis of the empirical research. Genet. Soc. Gen. Psychol. Monogr. **132**(1), 5–94 (2006)

R. Heslin, T.D. Nguyen, M.L. Nguyen, Meaning of touch: the case of touch from a stranger or same sex person. J. Nonverbal Behav. **7**, 147–157 (1983)

J. Hillman, *Emotion: a Comprehensive Phenomenology of Theories and Their Meaning for Therapy* (Northwestern University Press, Evanston, 1960)

D. Holman, R. Vertegaal, Organic user interfaces: designing computers in any way, shape or form. Commun. ACM **51**(6), 48–55 (2008)

J. Holt-Lunstad, W.A. Birmingham, K.C. Light, Influence of a 'Warm Touch' support enhancement intervention among married couples on ambulatory blood pressure, Oxytocin, Alpha Amylase, and Cortisol. Psychosom. Med. **70**(9), 976–985 (2008)

C. Hummels (1999) Engaging contexts to evoke experiences, in, *Proceedings of the conference Design and Emotion*, Delft University of Technology, pp. 39–45, 3–5 Nov 1999

H. Hutchinson et al, Technology probes: Inspiring design for and with families, in *Proceedings of CHI'03*, (ACM Press, Lauderdale, 2003)

C.E. Izard, J. Kagan, R.B. Zajonc, *Emotions, Cognition, and Behaviour* (Cambridge University Press, New York, 1984)

J.J. Jacobs (1999) How to teach, design, produce and sell product-related emotions, in ed. by C.J. Overbeeke, P. Hekkert *1st International Conference on Design and Emotion*, Delft University of Technology, pp. 9–14

K.W. Jacobs, F.E. Hustmyer, Effects of four psychological primary colors on GSR, heart rate and respiration rate. Percept. Mot. Skills **38**, 763–766 (1974)

J. James, R. Gross, W. Levenson, Hiding feelings: the acute effects of inhibiting negative and positive emotion. J. Abnorm. Psychol. **106**(1), 95–103 (1997)

D. Jeans, Changing formulation of the man-environment relationship in Anglo-American Geography. J. Geogr. **73**(3), 36–40 (1974)

R. Jensen, *The Dream Society* (McGraw Hill, New York, 1999)

C. Jones, M. Jonsson, Using Paralinguistic Cues in Speech to Recognise Emotions in Older Car Drivers, in ed. by C. Peterand, R. Beale, *Affect and Emotion in Human-Computer Interaction*. (Springer, Berlin, 2008) pp. 229–240

S.E. Jones, A.E. Yarbrough, A naturalistic study of the meaning of touch. Commun. Monogr. **52**, 19–56 (1985)

W.P. Jordan, *Pleasure with Products: Beyond Usability* (Taylor and Francis, London, 2000)

S.M. Jourard, An exploratory study of body-accessibility. British J. Soc. Clin. Psychol. **5**(3), 221–231 (1966)

J.W. Kalat, M. Shiota, *Emotion* (Wadsworth Publishing, Belmond, 2006)

J.E. Katz, *Machines That Become Us: The Social Context of Personal Communication Technology* (Transaction Publishers, New Brunswick, 2006)

H. Kawabata, S. Zeki, Neural corrolates with beauty. J. Neurophysiol. **91**, 699–1705 (2004)

E. Kennedy-Moore, J.C. Watson, *Expressing Emotion: Myths, Realities, and Therapeutic Strategies* (Guilford Press, New York, 1999)

J. Kim, E. André, Emotion recognition based on physiological changes in listening music. IEEE Trans. Pattern Anal. Mach. Intell. **30**(12), 2067–2083 (2008)

R.L. Klatzky, S.J. Lederman (2003) Touch, in ed. by A.F. Healy, R. Proctor, I.B. Weine, *Handbook of Psychology: Experimental Psychology 4*. Wiley, New York

M. Knapp, J. Hall, *Nonverbal Communication in Human Interaction* (Harcourt Brace College Publishers, New York, 1997)

K. Krippendorff, R. Butter, Product semantics: exploring the symbolic qualities of form innovation. J. Ind. Des. Soc. Am. **3**(2), 4–9 (1984)

D. Krish, *The Sentic Mouse: Developing a tool for Measuring Emotional Valence* (MIT (Department of Brain and Cognitive Sciences), Cambridge, 1997)

J. Lacan, *The Mirror Stage as Formative of the Function of the I as revealed in Psychoanalytic Experience*, (Trans. by A. Sheridan). (Tavistock, London, 1977)

C.L. Larson, J. Aronoff, J.J. Stearns, The shape of threat: simple geometric forms evoke rapid and sustained capture of attention. Emotion **7**, 526–534 (2007)

J.-P. Laurenceau, L.F. Barrett, P.R. Pietromonaco, Intimacy as an interpersonal process-the importance of self-disclosure, partner disclosure, and perceived partner responsiveness in interpersonal exchanges. J. Pers. Soc. Psychol. **74**(5), 1238–1251 (1998)

R.S. Lazarus, *Emotion and Adaptation* (Oxford University Press, New York, 1991)

H. Leder, C. Carbon, Dimensions in appreciation of car interior design. Appl. Cogn. Psychol. **19**, 603–618 (2005)

S.J. Lederman, R.L. Klatsky, Hand movements: a window into haptic object recognition. Cogn. Psychol. **19**, 342–368 (1987)

J. Lee, S. Marsella, Nonverbal Behavior Generator for Embodied Conversational Agents, in *6th International Conference on Intelligent Virtual Agents*, pp. 243–255 (2006)

J. Lee, J. Park, T.J. Nam, *Emotional Interaction Through Physical Movement* (HCI Intelligent Multimodal Interaction, 2007), pp. 401–410

B.I. Levy, Research into the psychological meaning of color. Am. J. Art Ther. **23**, 58–62 (1984)

K.C. Light, K.M. Grewen, J.A. Amico, More frequent partner hugs and higher oxytocin levels are liked to lower blood pressure and heart rate in premenopausal women. Biol. Psychol. **69**(1), 5–21 (2005)

L. Lipsenthal, Heart rate variability and emotional shifting: powerful tools for reducing Cardiovascular risk. News Heal. Health **5**(4), 2–4 (2004)

J. Liss, Central and peripheral mechanisms in the neurophysiology of depression. Hakomi Forum **5**, 18–24 (1987)

J.M. Loomis, S.J. Lederman, Tactual perception, in ed. by K. Boff, L. Kaufman, J. Thomas, *Handbook of Perception and Human Performance*. (Wiley, London, 1986) pp. 31–41

M. Merleau-Ponty, *Phenomenology of Perception*, Trans. by C. Smith (Routledge, London, 2005)

M. McLuhan, Understanding Media: The Extensions of Man (McGraw Hill, New York, 1964)

S. Marzano, A. Mendini, A. Branzi, *Television at the Crossroads* (Wiley, New York, 1995)

D. Matsumoto, American-Japanese cultural differences in the recognition of universal facial expressions. J. Cross Cult. Psychol. **23**, 72–84 (1992)

D. McDonagh-Philp, Gender and Design: Towards an appropriate research methodology, in *Proceedings of the 5th National Conference on Product Design Education July 1998.* (Glamorgan University, 1998)

B. Melin, U. Lundberg, A biopsychosocial approach to work-stress and musculoskeletal disorders. J. Psychophysiol. **11**(3), 238–247 (1997)

L.B. Meyer, *Emotion and Meaning in Music* (University Of Chicago Press, Chicago, 1961)

B. Mikellides, Color and physiological arousal. J. Archit. Plan. Res. **7**(1), 13–20 (1990)

A. Monk, M. Hassenzahl, M. Blythe, D. Reed, *Funology: Designing Enjoyment* (A Supplement to. Interactions, SIGCHI Bulletin, 2002)

A. Montagu, *Touching: The human significance of the skin* (Perennial Library, New York, 1986)

M.J. Moravcsik, *Musical Sound: An Introduction to the Physics of Music* (Plenum Publishers, New York, 2002)

M. Mori, The uncanny valley. Energy **7**, 33–35 (1970)

R. Morris, Crowdsourcing Workshop: The Emergence of Affective Crowdsourcing, in *CHI Workshop on Crowdsourcing and Human Computation*, (Vancouver, Canada), 7–12 May 2011

S. Mota, R. Picard, Automated Posture Analysis for Detecting Learner's Interest Level, in: *Workshop on Computer Vision and Pattern Recognition for Human-Computer Interaction*, CVPR HCI (2003)

M. Nagamachi, Kansei engineering: a new ergonomic consumer-orientated technology for consumer development. Int. J. Ind. Ergon. **15**, 3–11 (1995)

R. Neumann, B. Seibt, F. Strack, The influence of mood on the intensity of emotional responses: disentangling feeling and knowing. Cogn. Emot. **15**(6), 725–747 (2001)

M.L. Nguyen, R. Heslin, T.D. Nguyen, The meaning of touch: Sex and marital status differences. Represent. Res. Soc. Psychol. **7**, 13–18 (1976)

D.A. Norman, *Emotional Design: Why we Love (or hate) Everyday Things* (Basic Books, New York, 2004)

D.A. Norman, *The Design of Future Things* (Basic Books, New York, 2007)

M. O'Hagan, D. Ashworth, *Translation-mediated Communication in a Digital World: Facing the Challenges of Globalization and Localization* (Multilingual Matters, Clevedon, 2002)

K. Oatley, J.M. Jenkins, *Understanding Emotions* (Blackwell, Oxford, 1996)

C.J. Overbeeke, J.P. Djajadiningrat, C.C.M. Hummels, S.A.G. Wensveen, J.W. Frens, Let's make things engaging, in ed. by M.A. Blythe, K. Overbeeke, A.F. Monk, P.W. Wright, *Funology: From Usability to Enjoyment*. (Kluwer Academic Publishers, Dordrecht, 2003) pp. 7–17

V. Pallotta, Kinetic User Interfaces for unobtrusive interaction with mobile and ubiquitous systems, in ed. by D. Konstantas, J.M. Seigneur, *Mobile Quality of Service*. University of Geneve, 2009)

M. Pantic, A. Pentland, A. Nijholt, T.S. Huang, Human computing and machine understanding of human behavior: a survey artificial intelligence for Human computing, lecture notes. Artif. Intell. **4451**, 47–71 (2007)

A. Parkes, I. Poupyrev, H. Ishii, Designing kinetic interactions for organic user interfaces. Commun. ACM **51**(6), 58–65 (2008)

W.G. Parrott, J. Schulkin, Neuropsychology and the cognitive nature of the emotions. Cogn. Emot. **7**(1), 43–59 (1993)

M. Pavlova, A. Skolov, A.A. Skolov, Perceived dynamics of static images enables emotional attribution. Perception **34**, 1107–1116 (2005)

C. Peter, *Affect and Emotion in Human-Computer Interaction: From Theory to Applications* (Springer, New York, 2008)

C. Peter, E. Ebert, H. Beikirch, A Wearable Multi-Sensor System for Mobile Acquisition of Emotion-Related Physiological Data, in *Proceedings of the 1st International Conference on Affective Computing and Intelligent Interaction Beijing*, (Springer, Berlin 2005), pp. 691–698

Philips (2006) SKIN: Dresses. http://www.design.philips.com/philips/sites/philipsdesign/about/design/designportfolio/design_futures/dresses.page. Accessed 5 Jan 2013

R. Picard, *Affective Computing* (The MIT Press, Cambridge, 1997)

R. Plutchik, *A general psychoevolutionary theory of emotion* (Academic, New York, 1980)

F. Pollick, H.M. Patersona, A. Bruderlinc, A.J. Sanforda, Perceiving affect from arm movement. Cognition **82**(2), 51–61 (2001)

A. Rafaeli, I. Vilnai-Yavetz, Emotion as a connection of physical artifacts and organizations. Organ. Sci. **15**(6), 671–686 (2004)

V.S. Ramachandran, D. Brang, Tactile-emotion synesthesia. Neurocase **14**(5), 390–399 (2008)

M.L. Register, T.B. Henley, The phenomenology of intimacy. J. Soc. Pers. Relat. **9**(4), 467–481 (1992)

H.A. Ries, GSR and breathing amplitude related to emotional reactions to music. Psychon. Sci. **14**(2), 62–64 (1969)

B. Rimé, P. Philippot, S. Boca, B. Mesquita, Long-lasting cognitive and social consequences of emotion: Social sharing and rumination, in ed. by W. Stroebe, M. Hewstone, *European Review of Social Psychology 3*. Wiley, Chichester, 1992) pp. 225–258

P. Rodaway, *Sensuous Geographies: Body Sense and Place* (Routledge, London, 1994)

A. Rovers, H. van Essen, HIM: a framework for haptic instant messaging, in *CHI '04 Extended Abstracts on Human Factors in Computing Systems* (Vienna. ACM, New York, 2004)

J.A. Russell, A circumplex model of affect. J. Pers. Soc. Psychol. **39**(6), 1161–1178 (1980)

S. Planalp, *Communicating Emotion: Social, Moral, and Cultural*. (Cambridge University Press, Cambridge, 1999)

P. Salovey, M.A. Brackett, J.D. Mayer, *Emotional Intelligence: Key Readings on the Mayer and Salovey Model* (NPR Inc., New York, 2004)

S. Schachter, J.E. Singer, Cognitive, social, and physiological determinants of emotional state. Psychol. Rev. **69**, 379–399 (1962)

S. Schachter, *The Psychology of Affiliation*. (Stanford University Press, Palo Alto 1959)

K. Scherer, What are emotions and how can they be measured? Soc. Sci. Inf. **44**(4), 695–729 (2005)

H. Selye, Stress and disease. Science **122**, 625–663 (1955)

P.A. Shera, Selfish Passions and Artificial Desire: Rereading Clérambault's Study of "Silk Erotomania". J. Hist Sex. **18**(1), 158–178 (2009)

S.J. Lederman, R.L. Klatzky, Human Haptics, in ed. by L.R. Squire *New Encyclopedia of Neuroscience 5*, pp. 11–18, (2009)

C.A. Smith, P.C. Ellsworth, Patterns of cognitive appraisal in emotions. J. Pers. Soc. Psychol. **48**(4), 813–838 (1985)

M. Sonneveld, Dreamy hands: exploring tactile aesthetics in design, in ed. by D. McDonag, P. Hekkert, J. Erp, D. Gyi, *Design and Emotion 1*. (Taylor and Francis, London, 2004) pp. 228–232

A. Ståhl, K. Höök, M. Svensson, A.S. Taylor, M. Combetto, Experiencing the affective diary. Pers. Ubiquitous Comput. **13**(5), 365–378 (2008)

L. Stead, P. Goulev, C. Evans, E. Mamdani, The emotional wardrobe. Pers. Ubiquit. Comput. **8**(3–4), (2004)

J.C. Stevens, Thermal sensibility. in M.A. Heller, W. Schiff, *The Psychology of Touch*. (Erlbaum, Hillsdale, 1991) pp. 61–90

T. Schiphorst, Exhale: (breath between bodies), in *SIGGRAPH '05 ACM SIGGRAPH, Emerging Technologies*, Los Angeles, 2–4 Aug 2005

D. Tapscott, *Growing Up Digital* (McGraw-Hill, New York, 1997)

N. Tek-Jin, J.-H. Lee, P. Sun-Young Physical Movement As Design Element To Enhance Emotional Value of a Product. Design in *Proceedings of IASDR'07: International Association of Socities of Design Research—Emerging Trends in Design Research*, 12–15 Nov 2007, Hong Kong (2007)

S. Thayer, History and strategies of research on social touch. J. Nonverbal Behav. **10**, 12–28 (1986)

D. Tsetserukou, A. Neviarouskaya, Innovative real-time communication system with rich emotional and haptic channels. Lect. Notes Comput. Sci. **6191**, 306–313 (2010)

S. Turkle, *Life on the Screen: Identity in the Age of the Internet* (Simon and Schuster, New York, 1997)

P. Valdez, A. Mehrabian, Effects of color on emotions. J. Exp. Psychol. Gen. **123**(4), 394–409 (1994)

J.A.G.M. van Dijk, The Reality of Virtual Communities, in ed. by J. Groebel *Trends in Communication*. (Boom Publishers, Amsterdam 1997), pp. 39–63

T. Verhoef, C. Lisetti, F. Ortega, T. van der Zant, F. Cnossen (2009) Bio-sensing for emotional characterization without word labels, in Jacko JA (ed) Human-Computer Interaction, Part III, HCII 2009, LNCS 5612, pp. 693–702

J. Vincent, L. Fortunati, *Electronic Emotion: The Mediation of Emotion via Information and Communication Technologies* (International Academic Publishers, Oxford, 2009)

A.J.N. von Breemen, iCat: Experimenting with Animabotics, in *AISB 2005-Creative Robotics Symposium, 5 April 2005* Hatfield (2005)

E.H. Weber, *The Sense of Touch*. (Trans. by D.J. Murray). (Academic Press, New York, 1978)

G.D. Webster, C.G. Weir, Emotional responses to music: Interactive effects of mode, texture, and tempo. Motiv Emot. **29**, 19–39 (2005)

R. Webster, The cult of Lacan: Freud, Lacan and the mirror stage. http://www.richardwebster.net/thecultoflacan.html Accessed 5 Jan 2013 (1994)

S. Weinstein, Intensive and extensive aspects of tactile sensitivity as a function of body part, sex, and laterality, in D.R. Kenshalo, *The Skin Senses*. (Thomas, Springfield, 1968) pp. 195–222

S.J. Whitcher, J.D. Fisher, Multidimensional reaction to therapeutic touch in a hospital setting. J. Pers. Soc. Psychol. **37**, 87–96 (1979)

M.E. Yale, D.S. Messinger, A.B. Cobo-Lewis, C.F. Delgado, The temporal coordination of early infant communication. Dev. Psychol. **39**(5), 815–824 (2003)

R.B. Zajonc, On the primacy of affect. Amer Psychologist **39**, 117–123 (1984)

P. Zimmermann, S. Guttormsen, B. Danuser, P. Gomez, Affective computing-A rationale for measuring mood with mouse and keyboard. J. Occup. Saf. Ergon. **9**(4), 539–551 (2003)

Chapter 4
A Design Practice on Emotional Embodiment Through Wearable Technology

Abstract This chapter introduces the hypothesis, concept generation and the primary findings of the practice-based design research that aim at creating design tools, which serve to observe user behaviour towards the idea of communicating emotions through wearable technology. This research argues that emotions can be embodied through wearable technology in order to strengthen social bonds by adding more empathy and understanding to the human–human communication. In order to verify the hypothesis three concepts were developed: *Me&Myself, Nearby* and *Faraway*. Primary surveys on *somatic sensations of emotions* and the *usage of emotional wearable technology* were done in order to obtain useful data for building the prototypes that are based on three concepts.

4.1 Hypothesis

Clothes with their semiotic codes have always been used to express moods, sentiments or personality traits. However, clothes are less suited to express emotions, as emotions spontaneously change their state. Human skin is a surface, where emotions reveal themselves. Through the integration of textiles and technology, smart garments have begun to achieve almost all of the functions that skin can achieve. E-textiles can provide instant changes in the colour, pattern and form; therefore the garments become dynamic user interfaces that can display emotions. Today, wearable technology can find solutions for not only health care, but also social concerns that are rapidly changing due to the huge impact of communication technologies. Connectivity has become an important characteristic of the twenty-first century. The Internet has changed the structure of social interaction; shifting from constructed rigid social groups to fuzzily connected units. Personal privacy and intimacy has gained flexibility and new dimensions. Whilst society adapts to these new changes, design can introduce new concepts for human–human interaction. This research aims to create new communication agents that can be utilized as tools to observe user behaviour towards the idea of using wearable technology for communicating emotions. Emotional communication through wearable technology can strengthen social bonds and reconnect people in a way where emotions are not hidden anymore, but shown as tangible signals. Moreover, these signals can be transferred to other bodies in order to enforce the quality of communication over a long distance.

On the other hand, wearable technology, which observes vital body signals and reacts to them as a living artefact, can be perceived as an extension of the human body and, therefore can enable the wearer to embody emotions. The garments that can dynamically react to the emotional state of its wearer can lead to a strong personal attachment and feeling of ownership. Koskinen and Battarbee (2003) argue that living objects can grow attachment over time. The human body is the place, where emotional states are embodied as expressions or somatic sensations. Phenomenology argues that consciousness is embodied through the biological body and, furthermore the body is extended through artefacts. In this research, embodiment does not mean having a flesh that moves and talks; rather it means to extend the human body physically and cognitively into a new dimension that is shared with others. According to Merleau-Ponty (2005) *the lived body* (le *corps propre*), which experiences the awareness of having a body, cannot be separated from its own environment. Therefore, an embodied interaction can be possible by ignoring the limits of the raw body and extend these limits into a whole organism, which consists of all the artifacts that the human body is in contact with. While computer mediated communication is encouraging disembodiment, today with wearable technology it is possible to create an embodied interaction. While the body as a communication medium naturally exposes what goes on inside, today it is possible to transmit this information onto physical dynamic displays and by means of this, to other bodies so as to regenerate social bonds, which are constantly transforming with the influence of the new communication technologies. Emotional embodiment is strongly related with sensory systems, somatic sensations, expression and cognition. Therefore, this research focuses on how emotions can be transmitted by wearable technology that can simulate somatic sensations through multi-sensorial channels and what kind of cognitive results the wearable technology can elicit.

This research aims at finding new ways of the embodiment of emotions through wearable technology by extending sensorial experience beyond the body. This can bring more understanding and trust, and therefore positive influences to social interaction, obtained by technology. The hypothesis is tested and verified through an experimental design practice that is achieved by exploring, conceptualizing, building, testing and observing wearable tangible interfaces. According to Fishkin (2004), tangible interfaces can vary between two taxonomies—*embodiment* and *metaphor*—in order to create new user experiences. Embodiment is categorized as: *distant* (the output is separated from the input), *environmental* (output is around the user), *nearby* (the output is near to the input) and *full* (the output device is the input device). On the other hand, metaphor, in which the interface is related to another matter, is categorized as: *no metaphor* (no analogy), *noun or verb* (the interface looks like or acts like something else), *verb and noun* (the interface looks and acts like something else) and *full* (there is no distinction between the virtual and physical object) (Fishkin 2004). In this research the human skin is used as a *verbal* metaphor, in which the artefact acts like the human skin rather than mimicking its physical appearance. The research focuses on *distant, nearby* and *full* embodiment that provides a wide experimental area.

4.2 Concept Generation

The concept generation concentrates on creating new empathic communication tools for human–human interaction. Due to the fact that the research aims to enhance social interaction and support personal well-being, the concepts are developed according to two different interaction modalities. Emotion has significant roles in intrapersonal and interpersonal interaction. Interpersonal interaction involves others, while intrapersonal is a self interaction. An emotion can be helpful for a personal to make decisions, or it can enhance social bonds between two people. Based on this bi-dimensional role of emotion, concept generation consists of two interaction modalities: *intra* and *inter*. Intrapersonal interaction is the interaction between the user and the wearable agent (Fig. 4.1). The aim is to stimulate the sensorial system and help the wearer to be aware of his\her emotional states. It embraces a concept called *Me&Myself*.

On the other hand, interpersonal interaction is the interaction between the user and the others through the wearable agent. In this case, the aim is to communicate emotions to another person, who is nearby or distant. In the *Nearby* concept the wearable agent functions as a public display, while in the *Faraway* concept it can embody emotions of another person, who is at a far distance.

Although young generations are more adaptable and open to technological innovation, this research aims to find ways to connect people without any age difference. The research claims that this can be possible through engaging human body in the interaction that requires natural human behaviour rather than alienated mediation. The interaction incorporates human-based gestures and basic coding systems that can be easily learnt and applied. The embodiment of emotions through wearable technology aims at finding product solutions that can connect people with ease and create emphatic interaction.

Fig. 4.1 Three interaction concepts divided into two groups as inter-intra personal

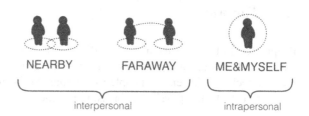

4.2.1 Me&Myself

The *Me&Myself* concept is an intrapersonal level of interaction, which occurs between the wearer and the wearable agent. Due to the fact that emotions are strongly related to psychological wellbeing, in this scenario emotions are communicated to the wearer as a feedback about their emotional state, especially to cope with stress. Stress is a common problem of today's lifestyle. Although it represents itself with various symptoms on the human body, it is very often not realized rapidly. Sometimes the person needs to understand his/her own feelings and analyse them in order to take care of her/his mental wellbeing. While daily life is becoming faster and faster, people cannot find time to focus on their own feelings and mental states. Most of the time people repeat the things that annoy them without realizing that their emotional state is being brought down. Stress directly affects the parasympathetic nervous system (Gleitman et al. 2004). Although this arousal makes the body alert, it can also cause stress-based exhaustion. This scenario envisions wearable agents that can inform the user about their inner state in order to obtain a somatic and mental awareness.

4.2.2 Nearby

Non-verbal emotional expression is essential for interpersonal relationships. People generally rely on the emotional expressions of the other people in order to give a relevant response (Darwin 1872). *Nearby* concept is based on wearable agents that are instantly changing shape according to the emotions of the wearer and attract attention of the nearby people. Due to the intimate substance of emotion, people sometimes may not prefer to express their emotions in public. Due to the context, a person can express, hide, or change his\her emotions (Kruml and Geddes 2000). Thus, this scenario is based on people who know each other and have an intimate relationship. Intimacy of emotions and social norms about emotional communication vary for different cultures, genders or environments, as explained in the background part of the research. Therefore, in this concept, product ideas are developed in order to analyse the user's perception on the issues of privacy and intimacy of communicating emotions through wearable technology.

4.2.3 Faraway

People generally prefer technologies that support and maintain social connections with friends, family and partners (Gibbs et al. 2005). Although digital communication technologies enable people to communicate with others that are at a long distance, it lacks the physical expression of emotions. *Faraway* concept is based on wearables that are instantly sending and receiving emotional data between two people, who are at a long distance. This concept is structured on different types of users and their needs regarding emotional communication.

Nowadays, elderly people are mostly detached from new technology, while younger generations are rapidly adapting to new forms of social interaction with new communication devices. The emotional wellbeing of elderly people can decrease because of being alone and out of reach from their families. Physical contact has many impacts in communication. Most elderly people live alone and they receive less physical contact than other people. Journard and Rubin (1968) argue that many people suffer from the lack of physical contact during their adult life. The literature infers that elderly generally receive "instrumental" touch rather than "expressive" touch (Watson 1975). Expressive touch can convey trust, hope and being connected with others. Hence, elderly people need to have more expressive touch.

On the other hand, mother and the child relationships can be another target for the *Faraway* concept. *Affectional bonding* is a type of attachment behaviour that can occur in the caregiver and his or her child (Bowlby 1979). Touch has very important influences in the attachment between mother and the child. Deep touch pressure can be applied as a therapy for autistic children to calm them down by an application of firm pressure (Grandin 1992).

Another user group that is targeted in this concept is long-distance relationships. Today with the growth of Internet usage these kinds of relationships are more sustainable. Technologies, such as cell phones, e-mail and video conferencing have made it possible to keep in touch with the distant partners. Due to the fact that these kinds of relationships require physical contact in order to communicate emotions, computer mediated communication cannot be sufficient enough to fulfil the needs of the partners.

Based on these diverse target groups, this concept focuses on product ideas, which enhance the tactile communication of emotions and bring alternative notions to visual and textual communication. It embraces products that can create tactile contact between two people in order to enhance their emotional attachment.

4.3 System Design

Emotion occurs in a process that embraces sensory system, motor system, arousal and cognition. Both the human body and mind is involved in this process. The human skin is a surface that is not only an actuator, but also a display. With its multi-layered structure it can sense and at the same time react to the stimuli. By taking the skin's function as a metaphor, a system is created in order to be a basis for the prototypes that are produced in the practice-based part of the research. The embodiment of emotions through wearable technology requires an interactive system, which consists of three main components: actuator, processor and display

(Fig. 4.2). The system works with input and output data. Actuators sense the input data, and then the processor processes this data and transmits it to the display, where the output appears as tactile, auditory or visual signals.

4.3.1 Actuator

The actuator works with three different inputs: vital body signals, gesture and self-activation in order to measure emotions. In this research the Russell's (1980) arousal-valance diagram is used in order to gather data that is obtained by the actuators.

Based on bio-monitoring studies, in the practice part of the research, HRV (Heart Rate Variability) and GSR (Galvanic Skin Response) sensors are used in order to measure the emotions of the wearer. GSR increases with arousal, while HRV may decrease with an appreciation or positive influence that is related to the valence level. These two parameters are gathered in order to measure the emotional state of the wearer by biosensors that are actuators of the system.

Besides the physiological measurement with biosensors, behavioural measurement is done by the use of sensors and mechanisms that track body movements and posture. Body posture and gestures are important non-verbal communication tools and they are mostly relevant to emotional states. In the practice-based part of the research these behavioural responses are used as parameters to measure emotions. This measurement is done by the use of basic pressure sensors, accelerometer and mechanisms that are triggered by human power.

Moreover, self-activation is a method of measuring emotions through self-reporting, giving the wearer freedom to indicate which emotions she/he feels instantly. In this modality, actuators work as on\off textile switches that activate the wearable system to express a basic emotion that is decided by the wearer.

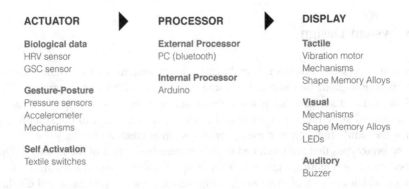

Fig. 4.2 System design that consists of actuator, processor and display

4.3.2 Display

Display is the surface, where the emotions are embodied. The display embodies the data, which comes from actuator in three modalities: tactile, auditory and visual. The research aims to create a predictable language in order to embody emotions on the display and avoid misinterpretations. This language is constructed as emotional embodiment coding patterns, which take the somatic and motor responses of the body (heartbeat and human movements) as basis.

The tactile expression of emotions is displayed in two ways: pressure and vibration. Pressure display is envisioned to mimic the tactile communication of emotions in human–human interaction. The more pressure is applied, the more arousal can be indicated. Vibration is another tactile modality that communicates emotions. It can simulate the rhythm of the heart beat -changing from irregular to regular- as an indicator of valence and the speed of the beat can indicate the arousal level of the emotion.

Visual display generates shape changes, movement and light in order to embody emotions. The dynamic wearables can shift from one state to another in order to express instant emotional changes. Due to the fact that people generally tend to find aesthetic objects more pleasurable, the aesthetic qualities of the wearables can transform from regular to irregular and can indicate emotions that show differences in the valance level. According to Aronoff et al. (1988), circular forms are mostly associated with positive emotions, and sharp corners are with negative emotions. The direction of the shape can represent emotions that have different arousal levels. If an object is out of it's direction, then it can give an impression of it's movement. Therefore, it may indicate an aroused emotion.

Movement is an important element of the display. Based on Laban's (1966) movement theory a movement coding system that has different variables (*flow* that is free or bound, *rhythm* that is irregular or regular, horizontal shape changes that are *spreading* or *enclosing*, vertical shape changes that are *rising* or *sinking* and time of movement that is *fast* or *slow)* is constructed in order to express emotions in different movement modalities.

Light is another display element that is used for embodying emotions. The application of light is similar to vibration. Lighting display simulates heartbeat with a change in rthym and speed. The inner rhythm of the human body is a strong element that can communicate emotions. This rhythm can be used as a regulator of psychological state. A dynamic display, which has variable rhythm of pulse, respiration or heart rate may affect biological responses and therefore can induce emotions. While rthym is the indicator of the valance, the speed corrolates with arousal. On the other hand, sound display is applied on the wearables as heartbeat sound that changes pitch, rythm and speed. These variablities are shown on the arousal-valance diagram in order to define emotional embodiment coding system (Fig. 4.3).

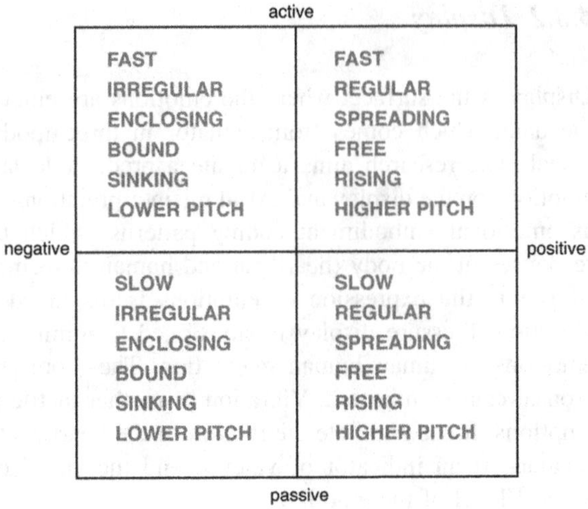

Fig. 4.3 Arousal-Valance diagram for speed, movement modality, rhythm and pitch

active

FAST IRREGULAR ENCLOSING BOUND SINKING LOWER PITCH	FAST REGULAR SPREADING FREE RISING HIGHER PITCH
SLOW IRREGULAR ENCLOSING BOUND SINKING LOWER PITCH	SLOW REGULAR SPREADING FREE RISING HIGHER PITCH

negative ——|—— positive

passive

4.3.3 Processor

The processor is programmed to convert the raw data based on the emotional embodiment coding patterns and sends an activation message to the display. It works as the brain of the human body that coordinates everything in the organism. In the practice part of the research two kinds of processors are used in order to merge and operate the data sent by actuators: external processor (PC and software) and internal micro-processor (Arduino). The external processor is the processor that is separated from the wearable agent and communication is achieved by bluetooth devices. On the other hand, the internal processor is embedded in the wearable agent with wired circuitry.

4.3.4 Data Transfer

Data transfer is another important aspect of the system. The architecture of the components (actuator, display and processor) can vary depending on how the data is transferred. The reciever and the sender are the main actors of the interaction, therefore the system architecture can change according to their role in the inter-action. For instance, in the *Me&Myself* concept the interaction happens between the wearer and the wearable agent, therefore the sender and the reciever are the same person. On the other hand, in *Faraway* concept the reciever and the sender are seperated people that are connected wirelessly.

4.4 Wearable Technology Prototyping

Due to the fact that wearable technology is a new field of study, there are some aspects that the designer should overcome. Wearable technology has two opposite sides: *wearing* that is related to softness and aesthetics; *technology* that is related to functionality. This contrast can be seen also in the prototyping process of wearable agents. Therefore, the designer should bridge these opposite aspects.

Textiles are soft; electronics are hard. This contradiction creates a difficulty that should be overcome by the designer. There are new materials available for building soft circuits and microprocessors, such as Lily pad that is embedded on the garment easily. However, in order to make the wearable agent function properly, there are other components with rigid structures that need to be embedded into the textile. Thus, the designer should position these components in the right places on the body in order not to irritate the user while moving, or avoid any problems of breaking. Garment building is a process mostly done by hand and relies on hand skills. However, smart technologies require knowledge of engineering and programming. The designer should have both skills that help her/him to create a better wearable agent. Generally, wearable technology projects are done in collaboration with different groups from different fields of research (varying from fashion to engineering). However, if the designer has those multi-disciplinary skills, this can enhance the level of the project and give rise to more innovative results. For normal clothes the aesthetic issue is very important in order to enhance the self-image. Therefore, the designer should bear in mind that wearable technology is not just a functional device attached to the human body; but a garment, which extends the wearer in society. Wearable technology should meet the aesthetical needs of an actual garment and bring human centred solutions to its additional functions through creating a balance between function and aesthetics. Moreover, wearable technology is a kind of converter that turns the digital data into tangible data. The designer should bear in mind this aspect of wearable technology and create new solutions, where the wearer can experience the digital data in a physical way with whole body involvement and consciousness.

The prototyping process consists of three parallel activities: electronics, garment making and interaction design (Fig. 4.4). These three activities are strongly connected to each other. For instance, the textile selection can affect the circuit

ELECTRONICS   GARMENT MAKING   INTERACTION DESIGN

Fig. 4.4 Prototyping process consists of three parallel activities: electronics, garment making and interaction design

building or; the sensors can define how data transfer is done. Therefore, the designer should work in parallel in these three areas of study and have knowledge about all of them.

4.5 Primary Findings

4.5.1 Survey on Somatic Sensation of Emotions

During emotional experience, every human being feels a unique somatic sensation. This data is important for the practice part of the research in order to define which kind of sensation has to be simulated through the wearable prototypes and where these prototypes should be located. To analyse the relation between emotions and somatic sensations, 100 participants (50 males and 50 females) selected from Europe, with an average age of 29, were asked to report how they experience six basic emotions (joy, surprise, sadness, fear, love and anger) in their bodies. A comprehensive list of somatic sensations and illustrations of the human body divided into different areas were given to the participants in order to answer the survey questions.

The first question was in which part of the body the participants feel the six basic emotions. Participants were asked to select the parts of the body on the given body illustrations (Fig. 4.5). The results showed that emotional experience occurs mostly (31 %) on the upper frontal part of the body, where the head, chest, shoulders and belly are located. Only 12 % of the participants chose the hands as the place where anger is experienced. The participants reported that love is felt mostly (44 %) on the chest, joy is felt mostly (22 %) in the mouth, fear is felt mostly (17 %) on the belly, sadness is felt mostly (14 %) on the shoulders and anger is felt mostly (38 %) on the upper head.

The second question was what kind of sensation the participants experience during the six basic emotions. The participants rated each emotion on a 10-point scale according to the duration of its somatic sensation (smooth or sudden) and they selected the type of sensation among the list, which was given in the beginning of the survey. Joy was reported as fullness (25 %), giggly sensation (24 %) and was rated as a smooth emotion (62 %) (Fig. 4.6). Surprise was reported as beating (22 %), bubbling (13 %), tense (10 %) sensation, and was rated as a sudden emotion (90 %) (Fig. 4.6). Sadness was reported as emptiness (30 %), pain (20 %), blocked (7 %) sensation, and rated as a smooth emotion (84 %) (Fig. 4.6). Fear was reported as tension (20 %), blocked (15 %), beating (14 %) sensation, and was rated as a sudden (75 %) emotion (Fig. 4.6). Love was reported as fullness (25 %), beating (15 %), bubbling (10 %) sensation and was rated as a smooth emotion (70 %) (Fig. 4.6). Anger was reported as tension (27 %), pressure (20 %), burning (16 %) sensation, and was rated as a sudden emotion (88 %) (Fig. 4.6).

Fig. 4.5   In which part of the body the participants (you) feel joy, surprise, sadness, fear, love and anger?

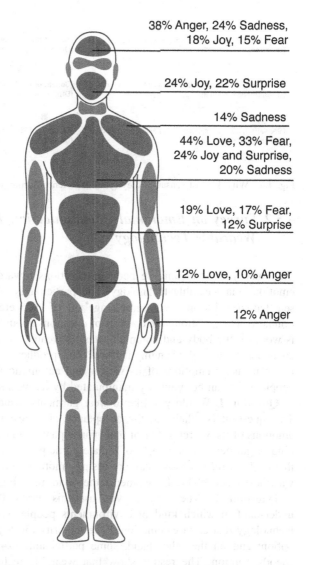

38% Anger, 24% Sadness, 18% Joy, 15% Fear

24% Joy, 22% Surprise

14% Sadness

44% Love, 33% Fear, 24% Joy and Surprise, 20% Sadness

19% Love, 17% Fear, 12% Surprise

12% Love, 10% Anger

12% Anger

These results show that there might be a common language for somatic sensation of emotions. For instance, in terms of duration, emotions like fear, anger and surprise are felt more suddenly, whereas love, sadness and joy are felt more smoothly. These results were used as basis for the design practice that consists of wearable agents that can simulate somatic sensations in order to transmit emotions.

JOY	SURPRISE
[Sensation] 25% fullness, 24% giggly, 18% bubbling [Duration] 62% smooth	[Sensation] 22% beating, 13% bubbling 10% tension [Duration] 90% sudden
ANGER	FEAR
[Sensation] 27% tension, 20% pressure, 16% burning [Duration] 88% sudden	[Sensation] 20% tension, 15% blocked, 14% beating [Duration] 75% sudden
SADNESS	LOVE
[Sensation] 30% emptyness, 20% pain, 7% blocked [Duration] 84% smooth	[Sensation] 25% fullness, 15% beating, 10% bubbling [Duration] 70% smooth

Fig. 4.6   What kind of sensation they (you) experience during joy, surprise and sadness?

4.5.2 Survey on Emotional Communication Through Wearable Technology

This survey aims to analyse people's reactions towards the idea of expressing emotions via wearable technology. 100 participants, 50 males and 50 females, selected from Europe, with an average age of 29, were asked to answer a questionnaire. The questionnaire consists of questions about the usage of a product that is worn on the body and communicates its wearer's emotions to a nearby or far away person through changing its form. This product can be used in two manners: it can measure emotions of its wearer and automatically exposes them to other people, or it can be manually manipulated by its wearer.

Question 1. Would you like to wear garments, which expose your emotions? This question is related to the acceptance of wearing technology that exposes emotions of its wearer. 80 % of male participants said that they would not use this kind of garments, while 54 % of female participants were positive about wearing them. These results show that expressing emotions through wearable technology was mostly acceptable for women rather than men (Fig. 4.7).

Question 2. Where would you use this garment? This question aims to understand in which kind of environments people would like to use wearable technology that exposes emotions. The majority (30 %) chose the "being alone" option; and on the other hand, some participants (24 %) chose the "bar with friends" option. The results show that wearable technology that expresses its wearer's emotions is mostly (30 %) considered to be used in a private environment (Fig. 4.7).

Question 3. Would you like to control your garment? This question aims at analysing the human behaviour towards emotional communication through wearable technology that can measure emotions and conveys the real inner state without self-masking. The answers show that the majority (72 %) of the participants chose to have self-controlled wearable agents instead of automatic ones. This result shows that people want to control what kind of affective information they express outside (Fig. 4.7).

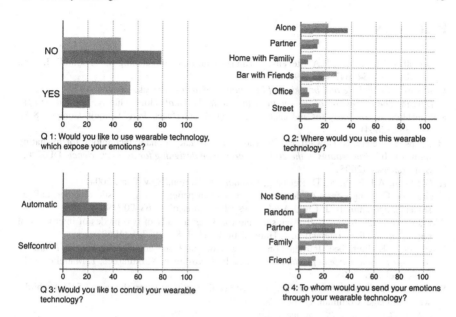

Fig. 4.7 Results of the survey on wearable technology and privacy (Light color indicates female, dark color indicates male)

Question 4. To whom would you send your emotions? This question aims to understand what kind of relationships can benefit from wearable technology that transmits emotions in a long distance communication scenario. 39 % of females and 24 % of males chose "partner" as an option to send their emotions through wearable agents. However, 40 % of the males wouldn't send the data. 22 % of females would like to send this emotional data to their families. These results show that females would like to send their emotions to people, with whom they have intimate relationships, such as, love partners and family members (Fig. 4.7). On the other hand, male participants who are interested in sending their emotions mostly (27 %) chose their partners that they have affective attachment.

The results of the survey show that females are more open to emotional communication through wearable technology than males. Wearable agents that dynamically communicate emotions are more accepted in a context, where the user is alone or with friends and partners. Public use is rarely considered acceptable. Females are more likely to send their emotions to a long distance person than males. Both genders prefer to control their garments with a self-activated system that gives them the freedom to decide which kind of emotion is sent or displayed.

References

J. Aronoff, A.M. Barclay, L.A. Stevenson, The recognition of threatening facial stimuli. J. Pers. Soc. Psychol. **54**, 647–655 (1988)

J. Bowlby, *The making and breaking of affectional bonds* (Tavistock, London, 1979)

C. Darwin, *The expression of the emotions in man and animals* (John Murray, London, 1872)

K.P. Fishkin, A taxonomy for and analysis of tangible interfaces. Pers. Ubiquit. Comput. **8**(5), 347–358 (2004)

M. Gibbs, F. Vetere, S. Howard, M. Bunyan, SynchroMate: a phatic technology for mediating intimacy. In: *Proceedings of the 2005 conference on designing for user experience*, DUX '05, San Francisco (2005)

H. Gleitman, A.J. Fridlund, D. Reisberg, *Psychology* (Norton, New York, 2004)

T. Grandin, Calming effects of deep touch pressure in patients with Autistic disorder, college students, and animals. J. Child Adolesc. Psychopharmacol. **2**, 63–70 (1992)

S.M. Journard, J.E. Rubin, Self-disclosure and touching: a study of two modes of interpersonal encounter and their interaction. J. Humanistic Psychol. **8**, 319–325 (1968)

I. Koskinen, K. Battarbee, Introduction to user experience and empathic design, in *Empathic design-user experience in product design*, ed. by I. Koskinen, K. Battarbee, T. Mattelmäki (IT Press, Helsinki, 2003)

S.M. Kruml, D. Geddes, Exploring the dimensions of emotional labor: the heart of Hochschild's work. Manage. Commun. Q. **14**, 8–49 (2000)

R. Laban, *The language of movement* (Plays Inc., Boston, 1966)

M. Merleau-Ponty, Phenomenology of perception. Translated by C. Smith Routledge, London (2005)

J.A. Russell, A circumplex model of affect. J. Pers. Soc. Psychol. **39**(6), 1161–1178 (1980)

W. Watson, The meaning of touch: Geriatric nursing. J. Commun. **25**, 104–112 (1975)

Chapter 5
Prototyping and Testing

Abstract In order to answer the research questions and verify the hypothesis that are mentioned in the previous chapter, virtual and physical prototypes were built and tested based on three interaction concepts: *Me&Myself, Nearby, Faraway*. The prototypes are the research tools, which can serve to observe user perception and behaviour towards the wearable agents that are designed according to these three concepts. While the virtual prototypes can be the initial tools to obtain primary results about user perception, the physical prototypes can provide more realistic results about user behaviours that are observed in real context with direct interaction with the prototypes.

5.1 Virtual Prototyping

Based on the *Nearby* concept, there are two characters that are in interaction: the wearer and the viewer. In this study, the virtual prototyping phase aims to test the viewer's perception towards a transformable garment that expresses the emotions of its wearer. A virtual dress was designed based on the Laban (1966) theory of movements in order to create a base for the study of real prototypes by analysing how emotions can be associated with different movements of the dress.

Folded and pleated textiles contain depth and volume; therefore they can give a dynamic appearance to a garment. The fold is a sign of action. In fashion design, pleats are often used as an element to emphasize the human form and add a dynamic look. Historically Roman drapery was used to represent movement (Hollander 1993). Like garments, the human skin also has folds. The natural pleats of human body, wrinkles, can appear when the person is expressing an emotion. An emotion can give rise to a tiny wrinkle that conveys a lot meaning more than the words can do. In the design phase of the virtual dress, the human skin was taken as a metaphor in order to create a dynamic garment that creases on different places of the body. The folds are placed on various parts of the dress and can move in two directions (horizontal and vertical) based on the Laban (1966) theory of

S. Uğur, *Wearing Embodied Emotions*, PoliMI SpringerBriefs,
DOI: 10.1007/978-88-470-5247-5_5, © The Author(s) 2013

movements. Horizontal folds on the dress occur for *spreading* and *enclosing* movements, vertical folds occur for *lengthening* and *shortening* movements. This study envisions that the *spreading* movement can express positive emotions; on the other hand *enclosing* movements can convey negative emotions. Depending on being vertical or horizontal, the arousal level of the emotion can also be indicated. The vertical folds can express more aroused emotions; on the other hand horizontal ones can express less aroused emotions. The dynamic dress was designed and constructed, by using a 3D-modeling tool in a virtual environment, in the VP Lab of KAEMART group (www.kaemart.it) in 2009. The 3D visualization tool was used to animate the changes on the dress. The animated 3D dress was attached on a real person's image in order to create a life-like video animation. While the person in the image was still, the dress was moving continuously in various modalities.

5.1.1 Virtual Dress Test and Results

The test was done in a laboratory setting and the animations of the dress were shown to the 10 participants (5 male, 5 female) on a computer screen. The participants were informed that the dress was communicating the emotional state of the person in the video. After the participants watched the animation, they reported their associations and impressions about what kind of emotion the wearer in the video expressed with the dress. The answers were analyzed in order to understand the relationship between the movement of the dress and the interpretations of the viewers (Fig. 5.1).

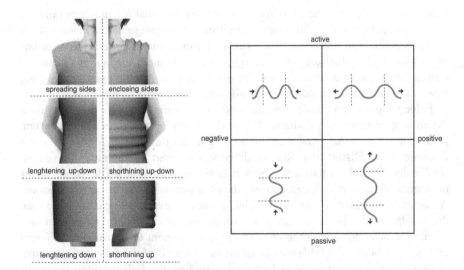

Fig. 5.1 Virtual dress and test results

Spreading to two sides was mostly (70 %) associated with positive emotions, being self-confident and energetic. Because of the spreading form, the dress was seen as a kind of growing shelter, which gives a stronger and bigger image to the human silhouette.

Enclosing from two sides was mostly (80 %) associated with fear and stress. The place of the folds also influenced the associations. Due to the fact that the folds were on the shoulder, it was mostly reported as muscle tension and therefore, stress. The participants reported that the dress was withdrawing itself, and they linked this movement mostly with fear.

Lengthening Up and Down was associated with relaxation. Because the movement looked like stretching, the participants mostly (60 %) associated it with peacefulness and being in a comfortable, relaxed psychological state.

Shortening Up and Down was associated with anxiety (30 %), depression (20 %) and disgust (50 %). Due to the fact that the folds were on the stomach area, most of the participants reported that the person had a problem that was related to the stomach, and therefore they made the association of being disgusted.

Lengthening Down—different than *Lengthening Up and Down*—was associated mostly (80 %) with sadness and depression.

Shortening Up—different than *Shortening Up and Down*—was associated with happiness (40 %), being energetic (20 %) and surprise (40 %).

While the vertical folds were associated with active emotions, such as anger, fear and stress, horizontal folds were associated with passive emotions, such as relaxation, disgust or depression. When the form change occurred in two directions—both vertical and horizontal,—most of the participants (80 %) were confused about the association of the emotions. These findings show that the direction of the movement could be an important element in order to express emotions. On the other hand, speed of the movement can give an idea about the arousal level of the emotion, such as fast closing movement can be associated with fear, and slow closing movement can be associated with depression. In this study, the speed is not considered as a variable; therefore all the movements of the dress were animated in the same speed that could be considered as neither fast nor slow.

5.2 Physical Prototyping: Social Skin

Based on the *Me&Myself* and *Nearby* concept, a set of product ideas—Social Skin—was designed in order to support interpersonal communication of emotions and also provide a self-awareness towards the inner state of the wearer. It functions as a dynamic display that communicates restrained inner states, such as stress, depression or excitement by tracking its wearer's biological data and posture. This study aims to analyse how people react to wearable technology that adds more transparency to interpersonal relationships and how both the wearer and viewer perceive these products. Social Skin prototypes are sensitive wearable agents that can measure the emotions of the wearer and change shape according to these measurements. The prototypes become parts of the human body, which warn its wearer about their inner states, such as stress that is harmful for the human body and enhance the communication between two people by conveying hidden feelings.

The prototyping and testing phase of Social Skin was realised at TU/e, Faculty of Industrial Design, Wearable Senses Laboratory in 2010. Two different modalities of measuring emotions (posture-based and biological) were used in order to track the wearer's emotions. While posture was measured by mechanisms that were trigged by muscle force, the biological data was obtained with body sensors (HRV and GSR sensors). The biological and postural data enable the wearable prototype to move and change shape. The prototypes were envisioned to move in four different modalities: shrinking, twisting, expanding and tapering. Each movement was obtained by different mechanisms, such as small servomotors or dynamic structures embedded into textiles. Although, Ni–Ti wire was inserted in some samples, in the final prototypes it was not preferred due to the fact that it reaches high temperatures and conducts high electrical current in order to become activated.

Social Skin embraces four wearable agents: Bubble-up, Fight-Flight, Under my Skin, and Skin&Bone. While Skin&Bone is a fully working prototype, the other three prototypes are conceptual and can partially function. These prototypes were constructed in order to test the wearer's and the viewer's perception towards the idea of dynamic garments that can instantly convey emotional messages. The user test was carried out firstly in a controlled environment, and secondly in a real-life condition. In the first user test the four primary prototypes were tested in order to analyze the viewer's perceptions towards the moving garments and Skin&Bone was selected as the most communicative wearable garment. Hence, a second test was done with Skin&Bone prototype in real-life setting in order to observe interaction patterns between the wearer and the viewer and it's influence on their interaction.

5.2.1 Bubble-Up

Bubble-up is a garment that is worn on the waist and expresses the excitement of the wearer (Fig. 5.2). The bubbles shift from a flat shape to an expanded shape when the arousal level increases. This prototype, as a primitive model was not improved by integrating sensors, because of incoherent results attained in perception studies. However, it was a practice exercise for exploring form and mechanism solutions to create dynamic movements on the garment.

The circular crochet pieces are attached to a cotton belt that covers the waist area. The crochet pieces can expand and become sphere like shapes by the movement of an inner mechanism that pulls and pushes the edges of the circles. The mechanism that has three circular metal wires is activated with a servomotor. With each cycle of the motor, the three circular metal wires get smaller and bigger, and hence, the crochet shapes can transform from flat shapes to bubble shapes.

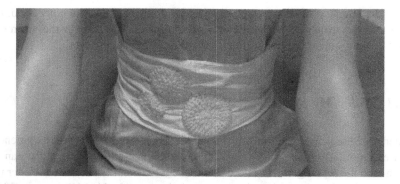

Fig. 5.2 Bubble Up prototype

5.2.2 *Fight-Flight*

Fight-Flight is a holster shape wearable prototype that can sense stress when the body posture becomes bent and tense (Fig. 5.3). When stress is detected, this textile structure responds by contracting. While the cone shape represents stress, the flat shape represents relaxation. In order to flatten the wearable garment, the wearer must make a more relaxed and healthy posture.

When the person is tense and stressed, the spine bends; therefore the shoulders also bend towards the chest. The holster exaggerates this posture and makes it more visible to other people. This wearable prototype uses the force that is created by muscle tension and converts it to a shape change in the textile structure attached on the holster. Transparent strings are embedded inside the holster and are attached to an armband. When the spine bends, the arm applies force to the strings and therefore, the strings activate the textile structure. Due to the fact that the communicative part of the wearable garment is on the backside of the wearer, it aims to

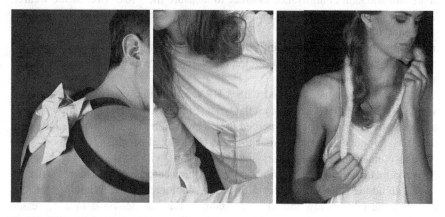

Fig. 5.3 Fight-Flight (*left*), Under my Skin (*middle*), Skin&Bone (*right*)

attract the attention of other people with a visual message. On the other hand, the force that the strings apply on the arm gives a tactile message to the wearer in order to warn him about his posture that indicates his stress level.

5.2.3 Under My Skin

Under my Skin is a product idea that changes shape when the wearer experiences aroused emotions (Fig. 5.3). This product is a kind of undergarment that can be attached to the actual garment in order to make it twist in the chest area. Under my Skin can sense happiness, joy and excitement by an embedded HR sensor. When Under my Skin is activated and moving, the actual garment moves, too. When the wearer feels excited and aroused, the wearable garment creates a twisting shape on the chest and communicates this feeling immediately to its wearer and to the people nearby. When the HR increases, the mechanism with a small servomotor starts twisting the garment. Although the prototype was built with Arduino (www.arduino.cc/) in order to make the servomotor move, it was not improved by the integration of a HR sensor.

5.2.4 Skin&Bone

Skin&Bone is a necklace that can measure the stress and depression level of its wearer and expresses this information with a shrinking movement (Fig. 5.3). The necklace consists of two different parts: a soft part, which acts like a dynamic skin moving around the neck; and a hard part, which contains the electronic components that give life to the necklace and keeps the soft part up. The necklace is part of a system, which uses the data of wireless body sensors. The sensors can monitor heart rate and skin conductance in order to measure the stress level of the wearer. When the wearer reaches to a certain level of stress, the soft part of the necklace starts moving up to the neck. While the shrinking movement can attract the attention of nearby people and inform them about the stress level of the wearer, by pulling the necklace down the wearer can become aware of her/his emotional state and therefore cope with stress.

Skin&Bone was accompanied with a HRV sensor that was worn on the chest and a GSC sensor that was worn on the wrist (Fig. 5.4). The two data streams were measured by body sensors and merged in a PC with software that gathered the real-time physiological signals in order to identify the level of stress in the body. The PC was connected to the sensors and the necklace with a Bluetooth connection. Skin conductance can give information about the wearer's arousal level, while the HRV can address the valance level of the psychological state. When the HRV is low, and the GSR is low too: this can indicate that the person is in a depressed emotional state. On the other hand, when the HRV is low, and the GSR is high:

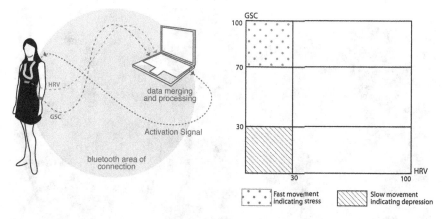

Fig. 5.4 Skin&Bone emotional measurement system

this can indicate that the person is stressed. Although, merging these two statistics can give reliable results about emotional states, the arousal and valance values can vary depending on personal properties, such as metabolism, gender, age, etc. Hence, the settings of the software were made according to average values of valance and arousal. When the software detects stress or depression, it sends a message wirelessly to the micro-controller inside the necklace to trigger the motor, which makes the soft part move. The necklace has its own movement language, which the viewer and the wearer should learn over time. For the stress mode the motor turns fast, and for the depressed mode it turns slowly.

The electronic components in the necklace were chosen to be as small as possible in order to put minimal weight on the neck. A small DC motor was used in order to pull and push the soft part of the necklace with a lithium battery as an energy source. The motor was triggered by a microprocessor that was linked to the software in the PC with Bluetooth. The soft part of the necklace was constructed with knitted wool.

5.2.5 Social Skin User Perception Test and Results

The user test was done in a laboratory setting, where the prototypes were placed on four mannequins (Fig. 5.5). Twelve participants (six male and six female) observed and interacted with the prototypes one by one. Each prototype was placed on different parts of the mannequin and could move in various modalities: shrinking, expanding, tapering and twisting. The participants were asked to report their perception towards each prototype by verbal and written forms of reports. The test was recorded with a video camera in order to observe the details of the interaction between the participants and the prototypes. The test results and observations were analysed based on: user interaction, movement, form and location.

Fig. 5.5 Social Skin user perception test

5.2.5.1 User Interaction

In the prototyping process, one of the most important factors was to design pleasurable, aesthetic and communicative products. Each material has different visual and tactile features. For instance, wool or felt can create more friendly touch experiences. In the perception test it was observed that the participants had a tendency to touch the moving prototypes due to their soft features. This tendency might also be explained by the curiosity of the movement, which invites them to touch. For instance, a participant expressed her opinion about Fight-Flight with these words:

> "It feels like he needs an attention like a tender touch or hug."

From the observations of the videos, it was seen that some of the participants (58 %) were mimicking the movement of the prototypes. For instance, when Under my Skin was twisting to left, most of the participants inclined their heads to the same direction. This observation shows that mirroring emotions can be seen in wearable technologies that can communicate emotion. Mirroring someone's emotions can create more lucid communication and increase intimacy. Therefore, these wearables might create an emphatic communication between two people.

The movement of the garment was a new thing for the participants; therefore it created excitement at the first gaze. 75 % of the participants reported excitement as a first impression that they experienced with the prototypes. The first impression that was caused by a primary appraisal could confuse the participants about the real emotional message that was given. The wearable's actual message could be perceived after a secondary appraisal with an association and cognitive process. The message that the prototype gives can be learned by time and, therefore the primary appraisal step can be skipped in the further trials.

The participants were asked which prototype was more effective in communicating an emotion and the results showed that Skin&Bone was the most (83 %) preferred one among the other prototypes. On the other hand, the mannequins were not real people; therefore the interaction was limited and could not simulate a real-time interaction. The results could be different, if the prototypes were worn by real people and in real conditions. Hence, this test gave an idea to do another test in real-time settings with Skin&Bone that was developed as a working model with body sensors.

5.2.5.2 Movement

According to the verbal and written form of the reports, the participants were mostly impressed with the movement of the prototypes. Due to the fact that the prototypes made unexpected movements, the participants associated them with an active emotion, such as excitement. Therefore, when the participants first spoke about their experience most of them (75 %) used the word 'excitement' due to the fact that they came across something new.

The user test also gave some results regarding motion perception. Four prototypes were simulating different movements (expanding, tapering, shrinking, twisting) at two different speeds (fast and slow). Although each movement creates different forms, in general they shift from a closed to an open state. Joy (33 %) and surprise (42 %) were linked with the fast opening movement. Relaxation (58 %) was linked with slow opening movement. Depression (42 %) was associated with a slow closing movement. Stress (58 %) and fear (33 %) were associated with fast closing movements.

5.2.5.3 Form

While the movement causes the first appraisal, the form is perceived and interpreted as a secondary appraisal. Therefore, associations and memory play a big role in the appraisal of an emotion with form perception. For instance, due to the fact that sharp objects are harmful, this kind of forms can be associated with negative emotions. When Fight-Flight became a thorn like shape, it was mostly (75 %) associated with a negative emotion, such as stress. When the prototypes appeared in aesthetically pleasant shapes, it was difficult for the participants to associate it with a negative emotion. In general, while disordered forms can communicate unpleasant emotions, symmetric and balanced forms can be associated with positive emotions (Overbeeke and Wensveen 2004). When the form shifts from order to disorder, this can be associated as a shift from pleasant to unpleasant emotion. For instance, Bubble-up prototype gave some confusing results, due to the disordered wrinkles in the textile. These wrinkles were associated with negative emotions, while the expanding bubbles were associated with positive emotions.

5.2.5.4 Location

The location of the prototypes on the mannequins effected the participants' perceptions, too. For instance, Fight-Flight that was placed on the shoulder and was associated with stress by 75 % of the participants. On the other hand, Bubble-up was placed on the stomach and was mostly (83 %) associated with excitement or anxiety.
Some of the participants reported:

It is like stress, when I have muscle tension on my shoulder.

Like a nervous feeling: when I am anxious, I feel it in my stomach.

To express their perceptions, the participants gave examples of the somatic sensations of the emotions and indicated the location that was effected by these emotions. The test showed that the location of the prototypes on the body could be a good indicator to express emotions.

5.2.6 Real Time Test with Skin&Bone and Results

According to the results of the previous test, Skin&Bone was found as the most communicative wearable garment. After it was completed as an advanced prototype working with body sensors, a real-time user test was done in order to understand what kind of the interaction patterns between the viewer and the wearer could emerge. The aim of this test is to see the influence of the prototype in real-time interaction. The communication of emotion between two people can only be unrestricted when they are in an intimate context and relationship. Therefore, a qualitative test was done in a real life environment with a couple, which are both designers and have been together for a long time, in order to provide wider and substantial information. The intimacy of the relationship gives the possibility to observe the role of the prototype in an intimate relationship, where it is easy to expose and express emotions.

The test was done in a real life environment without any recording in order to create a comfortable situation for the participants. They were left alone to have a daily conversation, while the prototype was measuring the female participant's biological data. After the test was completed, participants reported their one-hour experience. This report was analysed based on three issues: interaction, form and trust.

5.2.6.1 Interaction

There are two different interactions to be analysed: the interaction between the wearer and the necklace; and the interaction between the viewer and the wearer that is under the influence of the necklace. The wearer reported that it was comfortable to touch the necklace and pulling it down was relieving and even evoked relaxation. The movement of the necklace on the neck was merely perceptible as a tactile feedback, however the female participant could understand its movement from the changes of the male participant's facial expressions. In the beginning of the test session the necklace took-up a big part of their conversation. The main topic was the wearer's inner state and the necklace. However, later when they shifted to other topics, the necklace started to interfere with their conversation. Pulling the necklace down frequently distracted them from their topic.

5.2.6.2 Perception

In the test, Skin&Bone was considered as an aesthetically pleasant object that the female participant didn't have any problems to put on. She reported that it was pleasant for her to go around and she could wear it in a public environment. It did not seem like a weird technological device, which she would refuse to wear. However, the fact that it has some organic attributes, due to its movement, both participants perceived the necklace as a living object, an animal on their body. For instance, they reported their opinions:

The male participant: "It is like a snake trying to move around the neck."

The female participant: "It is like an animal that needs to be caressed."

Although it was seen as a living creature, the participants didn't show any negative feelings towards the necklace. In fact, they reported that the necklace was sympathetic to them and they had a tender feeling when they interacted with it.

5.2.6.3 Trust

Trust is another important issue of the Skin&Bone prototype, due to the fact that it captures biological data and automatically translates this data into movement. If the wrong body data combination triggers the wearable prototype, communication might fail and the necklace can give a different message than the real emotion. The participants reported that after one point they didn't take the necklace into consideration. They thought the necklace was moving without a justifiable reason. The fact that the necklace seemed to move without its' wearer's control, it might cause mistrust. The wearer reported that the necklace moved, when she was not experiencing stress. Therefore, both the wearer and viewer reported that in the end of the session they lost their trust and attention to the necklace. According to Picard (1997), person-dependent systems have different recognition rates than person-independent systems. The real-time user test showed that an uncontrollable wearable system might cause an unconvincing situation on the wearer. Therefore, in the next phase of the research, self-controlled systems that are activated by the wearer were built in order to continue testing the user perception.

5.3 Physical Prototyping: EMbody

EMbody is a set of product ideas that support the communication of emotions between two individuals, who are at a far distance based on the *Faraway* concept. The envisioned target group is people who have a strong affective attachment, but are in different places. Due to the fact that this kind of relationship requires mostly physical contact to express emotions, computer-mediated communication cannot be sufficient enough to fulfil the needs of communication. The envisioned wearable tools can instantly send and receive emotional messages in the form of tactile, auditory and visual signals by the will of the wearers. In this scenario, there are two wearers that send emotional messages to each other by gestural codes.

In order to build EMbody, two phases of prototyping were carried out. In the first practice, five different objects that give different sensorial outputs: vibration, light, sound, pressure and soft touch were developed. After these prototypes were tested by 20 participants, the test results oriented the research to create three different wearable agents, which were developed with more advanced prototyping techniques and tested in real-time communication settings by 12 participants.

5.3.1 Primary Prototypes

The results obtained in the survey *Somatic Sensation of Emotions* –fullness, emptiness, beating, pressure- are simulated by five different prototypes in three sensorial modalities: tactile, auditory and visual (Fig. 5.6). The first prototype is a soft object, which shelters a small vibration motor. This object can generate rhythmic vibrations that resemble a heartbeat when it is held inside the palms. It has four different patterns that require different rhythms and speeds: slow-irregular, fast-irregular, slow-regular and fast-regular. The aim of this prototype is to transmit heartbeat sensations via tactile stimulation. The second prototype is an armband that can simulate a pressure sensation by an inflatable mechanism covering the arm. When the bubble shaped elastic part expands, it applies pressure on the arm and fades out in few minutes. The third prototype is a soft object that applies a light touch with a mechanism that pulls and pushes points on the surface of the object. Due to its flexible and soft properties, it creates a tactile effect; as if the object is inhaling and

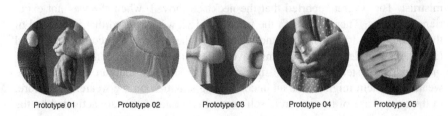

Prototype 01   Prototype 02   Prototype 03   Prototype 04   Prototype 05

Fig. 5.6 Five primary prototypes

exhaling that represent fullness and emptiness. In the fourth prototype, LEDs are embedded inside the soft surface in order to create four different blinking patterns with different rhythms and speeds: slow-irregular, slow-regular, fast-irregular and fast-regular. This prototype aims to create a heartbeat sensation via visual stimulation. The fifth prototype aims to create the same sensation via sound that changes in rhythm, pitch and speed. This soft object induces a low volume sound that directs the user towards an intimate interaction with the object.

Each prototype was built with the same neutral coloured textile and softness quality in order to create equal perception and concentration on sensorial stimuli. The prototypes were activated by mechanisms with small servo-motors, air pumped elastic surfaces, vibration motors, LEDs and buzzers.

5.3.2 User Test of the Primary Prototypes and Results

20 participants, 10 females and 10 males (with an average age of 27) attended the perception test. The results were obtained by written and verbal reports of the participants during and after the interaction. The test was done in a room with five objects on the table. The participants interacted with each object one by one, and after each interaction they reported their experience verbally. When they finished interacting, they reported their experience by answering specific questions and ranked it in order to have quantitative results. The results were analysed according to two issues: sensuous experience (physical) and negotiation (cognitive).

5.3.2.1 Sensuous Experience

The experience with the first object, which was simulating different heartbeat patterns as vibrations, showed that participants could easily recognize the arousal level with the speed of the beating, but the regularity of rhythm was not clear enough for them to identify the valence. For instance, irregular and fast beating was associated with anxiety, stress and fear (75 %). Besides, regular and fast heartbeat was associated mostly (55 %) with joy, happiness but also some of the participants (25 %) reported negative emotions, such as anxiety. On the other hand, slow regular vibrations were mostly (60 %) perceived as relaxation, but some of the participants reported negative emotions, such as sadness (15 %) or passive emotions (25 %) that are neither negative nor positive, like sleepiness. Irregular and slow vibrations were mostly (60 %) perceived as depression and sadness. In general, perceiving the arousal level was easier than the valence level. Mostly the participants were not sure about if the experience was positive or negative, although they could define the arousal level at the first moment.

Pressure was experienced with the second prototype that was filled with air. Most of the participants (75 %) associated this prototype with affection and they found it pleasurable. On the other hand, some participants (25 %) reported that the sensation was similar to negative tension. The fading out process was mostly (65 %) perceived as relaxation. The majority of the results show that deep touch pressure can induce positive emotions, such as relaxation. Although, the fading out process was mostly perceived as relaxation, some participants (35 %) reported that it was a kind of a sad feeling that they felt when they lost something, which belonged to them.

Light touch was simulated by the third prototype that had small points moving on the surface, creating inhale-exhale movement. This prototype simulated two tactile sensations: one was applied rapidly and repetitively; the other one was applied firmly and constantly. The rapid one was mostly (55 %) associated with fun and excitement. When it was activated firmly and constantly, it was mostly (75 %) associated with relaxation, closeness, and affection. Participants did not report any negative emotions. Participants had two different approaches while experiencing this prototype. They first watched the movement, and then they touched the moving object. When they visually experienced the object, they mostly (90 %) reported that the feeling was negative, such as being afraid or anxious. Some participants reported that the object seemed to withdraw itself as if it was afraid of something. On the other hand, when the participants put their hands on the surface, they experienced more positive emotions such as tenderness and a tickling like joy. Some of the participants found this movement relaxing as if it was massaging the hands.

Both the fourth and the fifth prototype simulate the heartbeat in two different modalities: visual and auditory. The results according to the rhythm and speed were quite similar with the first prototype. However, the participants reported that the first prototype was more effective.

The test results show that each sensorial experience has its own strengths and weaknesses. In order to get quantitative results, participants were asked to rank each prototype on a 0–10 scale according to their capacity of expressing emotions. The results show that the first prototype (6.2) and the second (5.3) got the highest rankings. Besides, the fourth prototype (4.4) was selected as third effective prototype. The third and the fifth prototype took the same rankings (3) and were found the least effective. The results of the rankings and oral reports show that tactile simulations can be more emphatic and effective in expressing emotions. Some of the participants reported that the heartbeat feeling that was simulated by the first prototype made them feel deeply affected. For instance, although the fourth and the fifth prototypes were doing the same rhythm and simulating the heartbeat, they described the tactile experience as more engaging. Most of them (65 %) reported that the first prototype could be an ideal way to communicate emotions. Besides, the participants generally (60 %) associated the fourth prototype with love and affection because of its lighting behaviour, although the rhythm could change this perception.

Participants had different interaction modalities for each prototype. For instance, while participants (60 %) stayed far from the fifth prototype that made sound, most of them (75 %) brought the first prototype near to their bodies during the interaction. The fourth prototype also created an intimate interaction; some participants (35 %) closed their two hands on the object in order to see the light more directly. They stated that they expected to have a warm sensation when they touched the blinking object; therefore they mostly held it with their hands as if they were putting their hands to the fire. The other observation was that the participants thought that the prototypes were interactive objects. Most of the participants (90 %) tried to change the state of the prototypes by shaking and squeezing and they searched for a button to control them. This shows that dynamic interfaces can create an expectation of one-to-one interactions.

5.3.2.2 Negotiation

After the participants reported the emotional associations that they made about each prototype, they were asked to explain the reason of these associations. They answered this question by using metaphors or analogies in order to explain how they felt. Their comments were mostly related to shape and material. The objects were associated with living things, such as human, animals or plants. For instance, some participants reported that the first prototype seemed them to be a small animal and the third prototype resembled to a creature that needed to be stroked. On the other hand, some participants reported that the second prototype was like a medical device and some of them related this object as a second skin on their arms. Some participants found the shape of the second prototype disgusting due to the fact that it seemed like an illness, which was growing on their arms. These results show that people tend to make associations and analogies when they come across with things that are new for them. Things that respond to human behaviours are mostly perceived as living objects; therefore they can create a meaningful attachment with their users.

5.3.3 EMbody: Final Prototypes

The final prototyping phase consisted of three different prototypes that were created in the light of the results obtained in the previous user test. Five modalities of emotional embodiment: vibration, pressure, light touch, sound and light were combined in three product ideas: *Sound Pad*, *Skin Deep* and *Hand-Muffs*. These prototypes were worn by two individuals that were willing to send their emotions to each other through hand gestures. These prototypes can translate the hand gestures into sensorial messages that communicate various emotions.

5.3.3.1 Hand-muffs

Hand-muffs are tactile soft wearable agents that can send and receive emotional hand gestures (Fig. 5.7). The idea is to combine the traditional hand-muff, which is

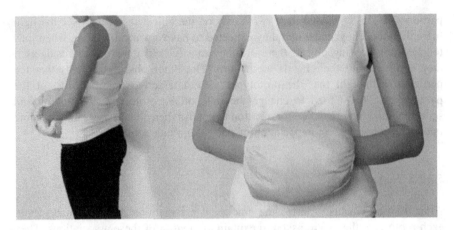

Fig. 5.7 Hand-muffs prototypes in use

used for warming hands in winter, with technology in order to create faraway emotional communication.

While deep touch pressure is a type of firm touch and can communicate affection and tenderness, light touch pressure requires less force and can express playfulness. Based on this, Hand-muffs were designed to simulate two depth levels of pressure: deep and light. Each muff has one touch sensor and a mechanism that is triggered by a small servomotor. The touch sensor of one muff can control the other muff's mechanism, controlled by Arduino. When one of the wearers activates the touch sensor placed inside the muff with rapid hand movements, the other wearer receives a playful light touch by fast closing-opening movements made by the mechanism embedded in the muff. On the other hand, when the wearer activates the touch sensor with constant pressure, then this is transmitted as a firm affective touch to the other person's muff by the continuous closing movement of the mechanism.

5.3.3.2 Sound Pads

Sound Pads worn by two people that are at a far distance is a set of shoulder pads, which can transmit emotional messages as auditory signals (Fig. 5.8). Small buzzers were embedded into the pads in order to generate beeping sounds, which are at a high and low pitch. The wearer can send each sound by touching the chest or shoulder. The chest sends a higher pitch beep that is envisioned as a positive emotion and shoulder sends a lower pitch beep that is envisioned as a negative emotion. The sound is received with a low volume in order to provide an intimate communication that can only be heard by the wearer. Textile buttons are attached to the chest and shoulder area and can control the buzzer of the other wearer's pad through the Arduino platform.

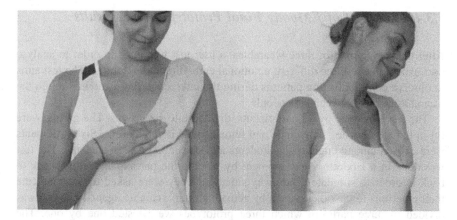

Fig. 5.8 Sound Pads prototypes in use

5.3.3.3 Skin Deep

Skin Deep is a set consisting of a collar and a bracelet that have vibration motors that simulate heartbeat sensation (Fig. 5.9). When one person touches his\her wearable, the other person receives the tactile message that is envisioned as the sender's heartbeat. While the collar provides the tactile sensation on the chest, where the heart is located; the bracelet provides it on the artery that is commonly used to measure the pulse. These wearables create an intimate communication, which allows the wearers to be in touch with their inner rhythms. Each prototype includes one touch sensor and vibration motor that are controlled by Arduino.

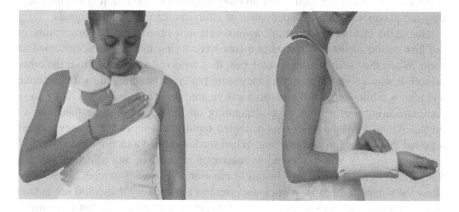

Fig. 5.9 Skin Deep prototypes in use

5.3.4 User Test of EMbody Final Prototypes and Results

After the prototyping of three wearables, a user test was done in order to analyse user experience in three different emotional embodiment modalities. The test aims to observe communication patterns during the interaction that is done through the wearable agents between two people.

The test was done by 12 participants, divided into six groups. The groups were set in female–female, male–male and female–male combinations. Six participants were chosen among engineering students and six among design students. The test was done in a laboratory environment by placing the participants sitting back to back in order to simulate a far away situation. They were asked to communicate their emotions to the other person by actuating their wearable agents. The test was divided in three parts, in which three prototypes were tested one by one. The participants filled a written form and reported their experience verbally, just after the test. The results and observations were analysed according to four issues: interaction, pleasure, sensorial experience and usage.

5.3.4.1 Interaction

Participants were left alone and asked to communicate with the other participants, who had the same wearable agent. They were asked to use the prototypes in order to send instant emotional messages to the other person. Due the fact that procedural knowledge can only be taught by demonstration and learned by practice (Norman 2002), some participants had problems with button indications and how to activate the prototypes. Therefore, the results show that these types of products need a demonstration on how to use.

It was observed that most (58 %) of the participants closed their eyes while interacting with the prototypes that gave auditory and tactile feedback. This shows that while receiving the tactile and auditory signals concentration is required and they are perceived with a deeper sense of attention than the visual signals.

Due to the fact that the wearable agents were new objects for the participants, in the first period of the test the participants were trying to become accustomed to them. When they activated the prototypes, they received a response from the other person. It was a sign for them that they were really communicating with someone and after a while some of the participants created their own language. They were communicating various emotions depending on the speed of their hand movements. When the Hand-Muff was activated rapidly and repetitively, they (58 %) reported that it was fun and exciting. When sender applied a tactile message firmly for a long time, it was perceived as relaxation, closeness, and affection (75 %). Participants didn't report any negative emotions. The environment would also effect the situation, because participants were in a laboratory that could not induce other emotions such as sadness or fear. Besides, it was observed that the participants, who were communicating with the Hand-Muffs, mimicked each other. For instance, when one made a rapid movement, the other one repeated it afterwards. When they received a message, they felt the need to answer, although they didn't really want to send an emotion.

Most of participants reported that the prototypes were not sufficient enough to give feedback whether their message was received or not. This situation caused ambiguity in their communication. The feedback was just the other person's tactile message. Therefore, the results show that when a person sends a message, this creates an expectation of an immediate reply. They reported that the curiosity of what the other person would send, made them excited and engaging.

5.3.4.2 Pleasure

In this test, Jordan's (2000) pleasure theory was used as a base of quantitative analysis. Participants ranked their experiences on a 0–10 scale according to four types of pleasure level: physio-pleasure that is related to the sensorial experience obtained from the experience, psycho-pleasure that is related to participant's psychological state during the experience, socio-pleasure that is related to the enhancement of the interaction with others and ideo-pleasure that is related to how others think about the wearer of the prototypes.

The rankings show that the prototypes can give socio and physio-pleasure more than the other types of pleasure (Fig. 5.10). Participants reported that the prototypes could increase their socio-pleasure, which support their engagement with others. Most of the participants reported that having sensorial feedback was pleasurable. On the other hand, ideo-pleasure, which is related to self-reflection, was given less ranking by the engineering students. They reported that wearing the prototypes in public was not pleasurable; moreover they would rather hide these prototypes under their own clothes. Most of the participants gave suggestions as having these objects in the form of normal clothes that were not recognizable as technological or weird. The students from design faculty had different rankings for the ideo-pleasure level of the experiment (Fig. 5.10). They found these objects to be pleasurable when worn in the public. Some of the participants reported that these objects could be "cool and fancy" to wear around. The results show that these kinds of products, which support an intimate communication of emotions, can enhance physio and socio pleasure of the user. Due to the fact that emotions are intimate, the users may prefer to communicate in an intimate way, which cannot be perceived by others.

5.3.4.3 Sensorial Perception

Each device induced different emotions in the participants. For instance, most of the participants (66 %) reported that the higher pitch-beeping signal obtained by Sound Pads was funny and joyful; on the other hand the lower pitch was annoying and associated with negative emotions. The firm pressure sensation obtained by

Fig. 5.10 Pleasure rankings
of the use experience

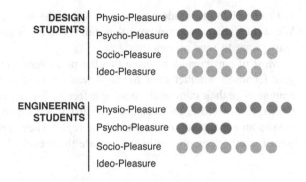

Hand Muffs was mostly (75 %) associated with relaxation, warmness and affection; the light touch was reported as playful as if it was tickling their hands by 58 % of the participants. On the other hand, the beating sensation in Skin Deep was mostly (66 %) associated with excitement and an aroused emotion. Participants reported they felt more intimately connected to the person with Skin Deep (66 %) and they could communicate more effectively with Hand Muffs (83 %). They reported that tactile based wearable prototypes were more effective in communicating emotions over a long distance.

5.3.4.4 Usage

The participants were asked in which kind of situations they would like to use these devices. The answers showed that participants mostly (83 %) preferred to use them in their intimate relationships, where words are not required anymore, but would like to send and receive affectionate messages. Most of the participants (66 %) stated that they would like to use the Hand Muff at home in a comfortable and intimate environment to concentrate and feel the intimacy. All female participants stated that they would use these devices with family or a partner; on the other hand male participants mostly (66 %) stated that they would use it with friends for fun. Some participants (25 %) stated that due to its colour and position on their arms, Hand-muffs resembled prostheses that could recall medical usage. On the other hand, some participants (33 %) stated that the Hand-muffs were like an animal or living thing on their knees that needed to be stroked; therefore they stated that it could be used as a calming object for the cure of psychological disorders.

References

A. Hollander, *Seeing Through Clothes* (University of California Press, Berkeley CA, 1993)
W.P. Jordan, *Pleasure with Products: Beyond Usability* (Taylor and Francis, London, 2000)

R. Laban, *The Language of Movement* (Plays Inc., Boston, 1966)

D.A. Norman, *The Design of Everyday Things* (Basic Books, New York, 2002)

C.J. Overbeeke, S.A.G. Wensveen, Beauty in use. Hum. Comput. Interact. **19**(4), 367–369 (2004)

R. Picard, *Affective Computing* (MIT Press, Cambridge MA, 1997)

Chapter 6
Conclusion

Abstract This chapter represents the reflections on the results of the user tests that were conducted to observe user behaviors towards the emotional wearable agents. The reflections underline various aspects of designing new interaction scenarios for wearable technology and question the power of technology over the human beings. This research provides a design methodology and knowledge that can be taken further and applied on real products that can meet with important user needs.

6.1 Interpretation of the Findings

Designers play an important role in pushing the limits of new technologies in order to answer users' needs that are constantly changing. Design refers to the act of creating. Therefore, a design point of view should bring an "action" to the research. This book addresses a practice based design research that translates the theoretical background into product ideas, which are used as research tools in order to analyze user perception and behavior towards the new applications of technology. Wearable technology has introduced new avenues for HCI through emphasizing the human body's role in interaction. This approach can also be implicated into technology-mediated communication in order to create more humanlike, emphatic interaction.

In the last decades, the DIY movement has gained importance in wearable technology due to emerging open source tools that provide huge knowledge about how to construct technologies that can be easily worn on the body with an aesthetic appearance. Wearable technology has shifted from being engineered to being designed by creative people and artists, who have skills to merge different kinds of knowledge. The research addressed in this book is an example of a multidisciplinary practice that involves people from different fields of study, such as engineers, and designers. The researcher (designer) is the main protagonist, who connects these different types of knowledge and form them with the light of aesthetics, ergonomics and human–human interaction. The researcher (designer) should have multidisciplinary skills and knowledge in order to create innovative solutions and cope with the complexities of the experimental study. The designer's role is to explore by experimenting and hence adding new knowledge to this new area of research.

This research explores paradoxical interactions between body and mind, inside and outside, near and far, private and public, natural and technological. One of these paradoxes is the soft and hard characteristic of wearable technology. It embraces textiles that are linked to aesthetics and fashion; and technology that is related to electronics and programming. It is still a new phenomenon for most people to use in their daily lives. While technology is getting smaller and smaller and invisibly woven into everyday objects, smart textiles and microelectronics have changed the general idea of wearing technology into a more aesthetic and subtle form. A smart garment can look like identical to a normal garment. However, it should be processed differently based on many aspects, such as weight, wearabilty, ergonomics, safety or assembling\disassembling, etc. Although human and technology integration is seen as cyborg, designers can turn this negative impression to a positive one by creating user-friendly products. The symbiosis of body and technology can be possible by functioning as an enhancement rather than a disturbance. To realize this, designers should know how the human body functions physically, psychologically and socially. Therefore, wearable technology should be designed in harmony with these three aspects that provide pleasure. As Jordan (2000) mentions, there is also another dimension of pleasure, which is called *ideo*. It is about how the user is perceived by others and how this situation can affect the user. For wearable technology this is a key issue. The studies on wearable technologies mostly try to answer the first three aspects of pleasure. While soft electronics can enhance physical pleasure by providing ergonomic solutions, biosensors can increase the psychological pleasure by measuring and assisting the emotional states of the wearer. Besides, integrating communication technologies into wearable technology can enhance social pleasure by creating instant connection with others. However, creating a look that is acceptable in society is sometimes missing. Therefore, the design approach should bring new ways of designing wearable technology that can be perceived as pleasurable to be worn in the public by bridging design and technology.

This research shows that technology worn on the body is not an unwanted phenomenon and people are open to new solutions to enhance their interactions with technology. However, they are concerned about looking strange or alien to others with these new kinds of technologies. Therefore, aesthetics of wearable technology is a very important aspect of the design process. Test results show that people want technology that does not look so different from their normal clothing, but can function as a smart object. The solution can be seen as designing actual clothes but with new materials. But, this would be an easy way to find a solution without going deeper and may cause negative side effects. At this point, a question appears: What is the difference between having a normal garment and a smart garment? In this research the prototypes are different than actual garments and have organic forms that were associated to living creatures by some of the participants in the user tests. They were perceived as extensions of the body. However, there is an important risk of creating an uncanny feeling that should be bared in mind. This research suggests that wearable technology should enhance the body image and functions; and the designer should not fall into the *uncanny valley* that can create negative effects on the perception of the technology that looks like and acts human. Therefore, this research tries to find new forms for dynamic wearables that can extend the human body in a more friendly way by avoiding misconceptions. Although this research argues that wearables should be extensions of the body, this idea doesn't support a prosthetic look but an aesthetic form of cohesion between body and technology.

On the other hand, emotion is embodied and experienced throughout the human body. To transmit the same emotion to another person is almost impossible. However, this research argues that with a measurement and simulating system this can be done. To measure emotions accurately is a very difficult task. As a matter of fact, measurement was the most challenging part of the research. The experiments show that there should be different measurement methods merged in one system to have more reliable results. However, this also might not give the exact idea about the emotions that the person feels. Therefore, this research also constructed a self-controlled mechanism in order to involve the wearer in the emotional measurement.

According to the research, touch is the most effective sensorial experience that can communicate emotions. Touch creates intimacy, attachment and a direct interaction. It can be sensed from everywhere on the body and convey various meanings depending on its properties. People can be attracted in front of a moving object but the tactile feeling of a moving object can affect them more strongly. Most of the emotions are experienced as different tactile sensations on the body, such as pain, pressure, heartbeat, burning, etc. Therefore, a tactile based communication system can work ideally in order to transmit emotions. Although tactile experience is an effective way of embodying emotions, an integration of sensorial experiences would be ideal, too. The case of Skin & Bone is an example of multimodal sensorial perception. The necklace has two properties: one is tactile that the wearer feels, and the other one is visual that the viewer sees. By having two characteristics, the necklace can communicate with both the wearer and the viewer about the wearer's current emotion. It can extend the tactile sensation to a second level that is visual.

On the other hand, emotions are also related to cognition that can be caused by negotiations. Form and colour are strong stimuli that elicit emotions and they can cause different associations and appraisals. For instance, the research found that the form and material could recall a past memory that effects the appraisal. Therefore, this research mostly focused on embodying emotion through simulating bodily sensations. The results pointed out that the simulations of sensations, such as deep heartbeats or breathing movement, could communicate emotions in a more efficient way. This shows that the human body can recognize and make sense of things that resemble its own substance, rather than things that are totally abstract.

This research shows that dynamic objects, which transmit emotions, can be considered as living objects and; therefore can attract people emotionally. This is strongly related to the cognitive level of perception that leads people to associate moving and responding objects as living mechanisms. This shows that dynamic wearables can create an emotional attachment with the user. On the other hand, the experiments done with haptic wearables showed that these prototypes could establish intimacy, attachment and a direct interaction, where technology as a physical entity dissolved from consciousness. People could concentrate on the interaction, rather than the wearable agents. This verifies that wearable technology can extend the limits of the body and consciousness by awakening sensorial perception. Besides, emotional communication through wearable technology can obtain constant attachment between two people. The theory of Deleuze and Felix (1988/1980), *Body without Organs* can be possible by connecting two people with an invisible tissue that creates a single body. From one side to another, this body can feel the same things. While using a tool, the perceptual experience is transferred to the end of the tool. The same perception can be created by haptic wearable technologies in order to extend the self to feel the emotions of the other person that is far away. Hence, the wearer can experience an ownership towards the wearable that is extraneous to his\her biological body. This research verifies that wearable technology that embodies emotions can enhance the social interaction, especially the intimate ones. For instance, the results of the tests show that people are more willing to use wearable technology to communicate with their intimate partners and families.

People express their emotions in order to be understood and they seek for empathy. Empathy can be possible by feeling the same way that another person feels. This practice based research tries to create empathic interactions between people by using technology as an invisible tool, rather than a main object. People sometimes cannot be expressive enough in emotional expression, or the other people cannot be empathetic enough to understand the person's intentions. Wearable technologies that can enhance emotional expression with their dynamic attitudes can create new social interaction patterns by increasing transparency and bringing lucidity. The *Nearby* concept was built on this issue to find product solutions in order to increase empathy between two people. However, the results show that people are not really comfortable about having this transparency in public environments.

Automatically activated garments are not desirable due to the fact that they could convey a message that the wearer doesn't want to convey. Rather, they want to control them and prefer garments that express the emotions that they want to express. This has similarities with the oldest function of clothing: covering. People cover their bodies to hide their skin. The skin is the surface that witnesses the deepest feelings of the human body and expresses them in various ways, such as blushing, shivering or sweating. By covering the skin, people somehow hide what needs to be expressed. The research shows that wearable technology has the same paradox that shifts from concealing to revealing. Automatically activated wearables can conceal the emotions that people don't want to express; therefore they prefer to control their garments. There is a considerable fact, social networks, where people have virtual personas and connect with others in virtual platforms.

While people hide their emotions in their real life, they can be frank in virtual platforms (Nardi et al. 2004). People would rather have digital distance while expressing emotions, which are not exposed in real life. This research tries to bring this phenomenon into real life, by putting the screens of computers away and adding intelligence to the screens of human body: garments. The results show that people are not ready to expose their inner state in public with wearable technologies that are dynamically interfering with their daily life interactions. On the other hand, they are more willing to express their emotions, which are mostly positive, such as tenderness, love, joy and excitement, to the people, with whom they have an intimate relationship.

While one of the roles of the garment is covering, wearable technology can make people naked in the sense of transparency of emotional expression. Clothing would just fade away, as if the wearable technology acts like a second skin. However, this research found that people would like to control this transmission of personal information. Therefore, a new question appears: "is wearable technology a new type of cover that gives people more power to hide under?" The answer of this question is up to the human will. Although technology seems as an unrestrained aspect that penetrates into the human life, people, who are the creators of the technology, can direct it according to their desires.

6.2 Contribution to Knowledge and Future Applications

This research provides a design methodology and knowledge that can be taken further and applied on user products that support emotional wellbeing. Although everyone needs to have this support, there are some specific areas that future research can be focused on. Aging population is a growing part of Europe and they are mostly out of reach from the online platforms that are used by majority of the young generation. Intel's Ethnographic Research on Aging Society reported that being socially isolated could cause negative effects on the elderly's state of wellbeing (Intel Corporation 2008). Hence, new product ideas based on the results of this research can be created to connect the elderly with their beloved ones, and provide physical interaction that can increase their emotional wellbeing. On the other hand, this research can be applied as product ideas for people, who are suffering from psychological disorders, in order to provide them with awareness of their inner state and regulate their emotions. For instance, fear, panic and anxiety disorders or mood disorders, like depression can be cured by therapeutic features of wearable technology. Besides, autism is another subject that can be an application area. Generally autistic people have difficulty to express their emotions and this can create problematic situations during the creation of social bonds. An autistic individual can stay calm during a highly aroused emotion (Picard 2009). Emotional embodiment through wearable technology can help people to regulate their emotions and communicate properly with others.

This research contributes to knowledge by providing a methodology of design practice for designers of wearable technology. It shows how a basic technology can give intelligence to garments in order to embody emotions by turning them into artificial skins extended from the human body over a near or far distance. While the way in which people experience emotions varies from one person to another, this research tries to find common language patterns in order to transmit emotions between two people for more empathetic and affective communication. By carrying out a design practice that consists of conceptualizing, building and testing processes, this study shows how a design practice can find new solutions to human needs that are strongly influenced by the technological improvements. In this sense, technology is not the core point of the innovation, but a mediator that supports the conceptual idea. The design practice proposes a new point of view that turns the negative side effects of technology into positive ones that may increase the social, psychological and physiological pleasure obtained by the use of technology. The research argues that the technology should not be seen as a tool, but an embedded intelligence that can give life to everyday objects and make them more emotionally responsive.

References

G. Deleuze, G. Felix. A Thousand Plateaus: Capitalism and Schizophrenia, vol. 2, Trans. Brian Massumi (London, Athlone, 1988/1980)

Intel Corporation Technology for an Aging Population: Intel's Global Research Initiative (2008), Available at: http://www.intel.com/Assets/PDF/general/health-318883001.pdf. Accessed 5 Jan 2013

W.P. Jordan, *Pleasure with Products: beyond Usability* (Taylor and Francis, London, 2000)

B. Nardi, D. Schiano, M. Gumbrecht, L. Swartz Why we blog. communications of the association for computing machinery. 41–46 Dec (2004)

R. Picard, Future affective technology for autism and emotion communication. Phil. Trans. Royal. Soc. B. **364**, 3575–3584 (2009)

Index

S. Uğur, *Wearing Embodied Emotions*, PoliMI SpringerBriefs,
DOI: 10.1007/978-88-470-5247-5, © The Author(s) 2013